Evaluation of the Sea Grant Program Review Process

Committee on the Evaluation of the Sea Grant Program Review Process

Ocean Studies Board

Division on Earth and Life Studies

NATIONAL RESEARCH COUNCIL
OF THE NATIONAL ACADEMIES

THE NATIONAL ACADEMIES PRESS
Washington, D.C.
www.nap.edu

THE NATIONAL ACADEMIES PRESS 500 Fifth Street, N.W. Washington, DC 20001

NOTICE: The project that is the subject of this report was approved by the Governing Board of the National Research Council, whose members are drawn from the councils of the National Academy of Sciences, the National Academy of Engineering, and the Institute of Medicine. The members of the committee responsible for the report were chosen for their special competences and with regard for appropriate balance.

This study was supported by Contract/Grant No. DG133R04CQ0009 between the National Academy of Sciences and the National Oceanic and Atmospheric Administration. Any opinions, findings, conclusions, or recommendations expressed in this publication are those of the author(s) and do not necessarily reflect the views of the organizations or agencies that provided support for the project.

International Standard book Number-10 0-309-10232-4
International Standard book Number-13 978-0-309-10232-2

Front cover: (Top) Elizabeth North checks her samples of blue crab larvae captured off the mouth of Chesapeake Bay. Photo credit: Michael W. Fincham. (Bottom) Photo taken during the Michigan Sea Grant Great Lakes Education Program (GLEP) benthic studies activity. Photo credit: Steve Stewart, Michigan Sea Grant College Program. (Right) Angie Sowers, Grace Brush and Holly Bowers (left to right) are taking core samples of sediment. Photo credit: Skip Brown. (Left) Marine biologists from Point Reyes Bird Observatory (PRBO) rinsing nets after capturing krill at the shelf-break near California's Farallon Islands. Photo credit: Benjamin L. Saenz, PRBO. *Back cover:* (Top) Charleston Progressive middle-school students bag shell on Bowens Island for the S.C. Oyster Restoration and Enhancement Program. Photo courtesy of S.C. Department of Natural Resources. (Right) Stephen Giovannoni shown studying harmful algae that affect shellfish. Photo courtesy of Oregon Sea Grant. (Left) Teachers learn to build their own remotely operated vehicle during the Sea Perch workshop. Photo courtesy of S.C. Sea Grant Consortium.

Additional copies of this report are available from the National Academies Press, 500 Fifth Street, N.W., Lockbox 285, Washington, DC 20055; (800) 624-6242 or (202) 334-3313 (in the Washington metropolitan area); Internet, http://www.nap.edu.

THE NATIONAL ACADEMIES

Advisers to the Nation on Science, Engineering, and Medicine

The **National Academy of Sciences** is a private, nonprofit, self-perpetuating society of distinguished scholars engaged in scientific and engineering research, dedicated to the furtherance of science and technology and to their use for the general welfare. Upon the authority of the charter granted to it by the Congress in 1863, the Academy has a mandate that requires it to advise the federal government on scientific and technical matters. Dr. Ralph J. Cicerone is president of the National Academy of Sciences.

The **National Academy of Engineering** was established in 1964, under the charter of the National Academy of Sciences, as a parallel organization of outstanding engineers. It is autonomous in its administration and in the selection of its members, sharing with the National Academy of Sciences the responsibility for advising the federal government. The National Academy of Engineering also sponsors engineering programs aimed at meeting national needs, encourages education and research, and recognizes the superior achievements of engineers. Dr. Wm. A. Wulf is president of the National Academy of Engineering.

The **Institute of Medicine** was established in 1970 by the National Academy of Sciences to secure the services of eminent members of appropriate professions in the examination of policy matters pertaining to the health of the public. The Institute acts under the responsibility given to the National Academy of Sciences by its congressional charter to be an adviser to the federal government and, upon its own initiative, to identify issues of medical care, research, and education. Dr. Harvey V. Fineberg is president of the Institute of Medicine.

The **National Research Council** was organized by the National Academy of Sciences in 1916 to associate the broad community of science and technology with the Academy's purposes of furthering knowledge and advising the federal government. Functioning in accordance with general policies determined by the Academy, the Council has become the principal operating agency of both the National Academy of Sciences and the National Academy of Engineering in providing services to the government, the public, and the scientific and engineering communities. The Council is administered jointly by both Academies and the Institute of Medicine. Dr. Ralph J. Cicerone and Dr. Wm. A. Wulf are chair and vice chair, respectively, of the National Research Council.

www.national-academies.org

*This report is dedicated
to the memory of committee member
Fred Noel Spiess (1919-2006)*

COMMITTEE ON THE EVALUATION OF THE SEA GRANT PROGRAM REVIEW PROCESS[1]

JAMES M. COLEMAN (*Chair*), Louisiana State University, Baton Rouge
ROBERT J. BAILEY, Oregon Department of Land Conservation and Management, Salem
BILLY J. (B.J.) COPELAND, North Carolina State University, Pittsboro
SUSAN E. COZZENS, Georgia Institute of Technology, Atlanta
KEITH R. CRIDDLE, University of Alaska, Fairbanks
ELIEZER GEISLER, Illinois Institute of Technology, Chicago
MICHAEL W. HOWELL, University of South Florida, St. Petersburg
RICHARD C. KARNEY, Martha's Vineyard Shellfish Group, Inc., Vineyard Haven, Massachusetts
GEORGE I. MATSUMOTO, Monterey Bay Aquarium Research Institute, Moss Landing, California
JOAN BRAY ROSE, Michigan State University, East Lansing
ANDREW R. SOLOW, Woods Hole Oceanographic Institution, Woods Hole, Massachusetts
FRED N. SPIESS, Scripps Institution of Oceanography, University of California, La Jolla

Staff

DAN WALKER, Study Director
JENNIFER MERRILL, Senior Program Officer
AMANDA BABSON, Christine Mirzayan Science and Technology Policy Fellow
NANCY CAPUTO, Research Associate

---

[1]Committee and staff biographies can be found in Appendix A.

Acknowledgments

This report was greatly enhanced by the participants of the three workshops held as part of this study. The committee would first like to acknowledge the efforts of those who gave presentations at meetings: Ronald Baird, NOAA; Leon M. Cammen, NOAA; Jonathan Kramer, Sea Grant Association; Frank Kudrna, Sea Grant Review Panel; Nathaniel E. Robinson, Sea Grant Review Panel; and Francis Schuler, NOAA. These talks helped set the stage for fruitful discussions in the closed sessions that followed.

The committee is also grateful to a number of people who provided important discussion and/or material for this report, these include:

Brian Allee, Alaska Sea Grant, University of Alaska Fairbanks
Paul Anderson, Maine Sea Grant, University of Maine, Orono
Anders W. Andren, Wisconsin Sea Grant, University of Wisconsin-Madison
James C. Cato, Florida Sea Grant, University of Florida, Gainesville
Ruperto Chapparo, Puerto Rico Sea Grant, University of Puerto Rico, Mayaguez
Barry A. Costa-Pierce, Rhode Island Sea Grant, University of Rhode Island, Narragansett
Penelope D. Dalton, Washington Sea Grant, University of Washington, Seattle
M. Richard DeVoe, South Carolina Sea Grant Consortium, Charleston

Linda E. Duguay, University of Southern California Sea Grant Program, Los Angeles
Louie Echols, Washington Sea Grant, University of Washington, Seattle
Lawrence K. Forcier, Lake Champlain Sea Grant, University of Vermont, Burlington
Nikola Garber, NOAA, Silver Spring, Maryland
E. Gordon Grau, Hawaii Sea Grant, University of Hawaii, Honolulu
Sami Grimes, NOAA, Silver Spring, Maryland
Jeff Gunderson, Minnesota Sea Grant, University of Minnesota, Duluth
Ronald Hodson, North Carolina Sea Grant, North Carolina State University, Raleigh
Mark Malchoff, Lake Champlain Sea Grant, University of Vermont, Burlington
Robert Malouf, Oregon Sea Grant, Oregon State University, Corvallis
Jack S. Mattice, New York Sea Grant, State University of New York, Stony Brook
Judith E. McDowell, Woods Hole Oceanographic Institution Sea Grant, Woods Hole, Massachusetts
Russell A. Moll, California Sea Grant, University of California-San Diego, La Jolla
Edward C. Monahan, Connecticut Sea Grant, University of Connecticut, Groton
Amy Painter, NOAA, Silver Spring, Maryland
Jonathan Pennock, New Hampshire Sea Grant, University of New Hampshire, Durham
Melissa Pearson, NOAA, Silver Spring, Maryland
Mac V. Rawson, Georgia Sea Grant, University of Georgia, Athens
Jeffrey M. Reutter, Ohio Sea Grant, Ohio State University, Columbus
Carl Richards, Minnesota Sea Grant, University of Minnesota, Duluth
William L. Rickards, Virginia Sea Grant, University of Virginia, Charlottesville
Donald Scavia, Michigan Sea Grant, Ann Arbor
Robert R. Stickney, Texas Sea Grant, Texas A&M University, College Station
William Sullivan, Illinois-Indiana Sea Grant, University of Illinois Urbana-Champaign

LaDon Swann, Mississippi-Alabama Sea Grant Consortium, Ocean Springs, Mississippi
Nancy Targett, Delaware Sea Grant, University of Delaware, Newark
Michael P. Weinstein, New Jersey Sea Grant, New Jersey Marine Science Consortium, Fort Hancock
Charles Wilson, Louisiana Sea Grant, Louisiana State University, Baton Rouge

This report has been reviewed in draft form by individuals chosen for their diverse perspectives and technical expertise, in accordance with procedures approved by the NRC's Report Review Committee. The purpose of this independent review is to provide candid and critical comments that will assist the institution in making its published report as sound as possible and to ensure that the report meets institutional standards for objectivity, evidence, and responsiveness to the study charge. The review comments and draft manuscript remain confidential to protect the integrity of the deliberative process. We wish to thank the following individuals for their participation in their review of this report:

Robert A. Duce, Texas A&M University, College Station
Dennis Hedgecock, University of Southern California, Los Angeles
Susan M. Henrichs, University of Alaska Fairbanks, Fairbanks
Ronald N. Kostoff, Office of Naval Research, Arlington, Virginia
Jonathan Kramer, University System of Maryland, College Park
Christopher G. Mann, Marine Aquaculture Task Force, Takoma Park, Maryland
Bonnie J. McCay, Cook College of Rutgers University, New Brunswick, New Jersey
Arthur R. Nowell, University of Washington, Seattle

Although the reviewers listed above have provided many constructive comments and suggestions, they were not asked to endorse the conclusions or recommendations nor did they see the final draft of the report before its release. The review of this report was overseen by **Michael K. Orbach**, Duke University, appointed by the Divison on Earth and Life Studies, who was responsible for making certain that an independent examination of this report was carried out in accordance with institutional procedures and that all review comments were carefully considered. Responsibility for the final content of this report rests entirely with the authoring committee and the institution.

Contents

Summary

The National Sea Grant College Program (NSGCP or Sea Grant) was created nearly 40 years ago and has matured into a state-federal partnership with a distinctive role and management structure. Sea Grant is a nationwide network (administered through the National Oceanic and Atmospheric Administration [NOAA]) of 30[1] individual Sea Grant programs[2] based at some of the nation's top universities. The NSGCP engages this network in conducting scientific research, education, training, and extension projects designed to increase assessment, development, utilization, and conservation of coastal resources by providing assistance to promote responsive research and training activities and to broaden knowledge and techniques (National Sea Grant College and Program Act, 1966 [P.L. 89–688]).[3]

The NSGCP has been a main source of funding in the United States for activities in marine policy, and thus far has been a major contributor to the issues of aquaculture, biotechnology, coastal communities and economies, coastal natural hazards, ecosystems and habitats, fisheries, marine science literacy, seafood science and technology, urban coasts and invasive species. The program also supports students at all levels of the edu-

[1]Not including the 3 programs in development stages.

[2]For the purpose of this report, all 30 programs will be referred to as "individual Sea Grant programs." Previous Sea Grant literature has also used the term "state program" or "Sea Grant college/institute."

[3]See Appendix C for Sea Grant legislation.

cational system and has supported education and training of many marine and Great Lakes scientists, resource managers, and policy specialists through its three fellowship programs, including the John A. Knauss Marine Policy Fellowship, the Sea Grant/NOAA Fisheries Graduate Fellowship, and the Sea Grant Industry Fellowship Program.

In 1993, the Under Secretary of Commerce for Oceans and Atmosphere requested the National Academies review and evaluate the NSGCP as part of an effort to prepare for the then pending National Sea Grant College Program Reauthorization Act of 1998 (P.L. 105-160). The resulting 1994 report, *A Review of the NOAA National Sea Grant College Program,* recommended several actions, including strengthening the strategic planning process at the national level, clarifying the roles and responsibilities of the National Sea Grant Office (NSGO) and individual program personnel as well as the National Sea Grant Review Panel (NSGRP), and carrying out systematic, periodic reviews of the individual programs (National Research Council [NRC], 1994).

THE CURRENT STUDY

In partial response to the 1994 report, the Director of the NSGCP (referred to as "National Director" throughout this report) requested that the National Sea Grant Review Panel establish a process for evaluating each individual program once over a four-year review cycle. These reviews are carried out through a series of site visits, each of which usually involves 4 to 7 recognized experts in marine science and policy, who focus on a uniform set of performance criteria, using a standardized set of benchmarks and indicators. This evaluation process has evolved through time, in response both to experience gained during its execution and to evolving expectations of Congress. The National Sea Grant College Program Act Amendments of 2002 (P.L. 107–299) directed NOAA to contract with the National Academies to carry out a review of the evaluation process and make appropriate recommendations to improve its overall effectiveness.

Statement of Task

The Committee on the Evaluation of the Sea Grant Program Review Process (the Committee) was charged with assessing new procedures adopted by the NSGCP since the publication of the 1994 NRC report to determine their impacts. During this study, the Committee assessed the impact of the new procedures and evaluation process on Sea Grant as a whole. Among the areas considered were the quality of the work produced by the program; its responsiveness to national, regional, and local needs; and the quality of its leadership, management, and reputation. Specifically, the Committee was asked to examine:

(1) Effectiveness of major changes instituted in response to the recommendations of the 1994 NRC report with regard to individual program performance and quality.

(2) Effectiveness of program review procedures with regard to accuracy, accountability, and enhancement of individual program performance. Both the previous and current review procedures (adopted in 2003 in response to the Sea Grant Act of 2002) will be assessed as specified below:

- Review the effectiveness of the evaluation and rating system in determining relative performance of programs with regard to management and quality of research, education, extension, and training activities;
- Evaluate whether there have been improvements in programs as a result of the evaluation process;
- Evaluate the 2003 review procedures for their ability to meaningfully segregate individual programs into five categories based on competitive scores; and
- Compare the effectiveness of the previous and 2003 review procedures with regard to the dual objectives of maximizing the quality of each program and of rating programs relative to each other for the purpose of determining performance-based funding.

(3) Assessment of the usefulness and fairness of metrics developed to evaluate programs with different operational constraints, resources, and local priorities.

- Evaluate metrics for relevance and clarity;
- Determine whether metrics provide a quantitative measure of quality of performance; and
- Assess whether metrics improve consistency and objectivity of reviews from different teams evaluating a diverse portfolio of individual Sea Grant programs.

The Committee was also asked to make recommendations for improving the overall effectiveness of the evaluation process to ensure fairness, consistency, and enhancement of performance.

IMPACT OF CHANGES IN RESPONSE TO 1994 REPORT

Following the 1994 NRC report, the NSGO instituted a number of changes in an effort to improve the overall program and the manner in which individual programs are evaluated. Although strategic planning within the NSGCP needs to be improved, the adoption of a formal strategic planning process at the national program level, as recommended in the 1994 report, is *prima facie* an improvement over earlier practice. In addition, there is a consensus among the directors of individual Sea Grant programs that the evaluation process instituted in 1998 in partial response to the 1994 report has led to improvements in their programs, despite the fact that many within this group are openly critical of some aspects of the process. Finally, several members of the Committee have first-hand, long-term experience with the Sea Grant program and it is their considered opinion that the changes instituted since 1994 have strengthened the overall program. As with the Sea Grant directors, the opinions of even knowledgeable individuals cannot be taken as objective indicators; but, the unanimity of response to this issue—particularly in light of differences of opinions on other issues—suggests that real improvements have occurred.

EFFECTIVENESS OF POST-2002 EVALUATION

As mentioned, the process established by the NSGO in 1998 (and modified periodically since) to evaluate program performance appears to have led to improvements in the overall program. However, several areas of concern remain. Since the reauthorization of the program in 2002, program evaluation within Sea Grant has evolved to serve two, theoretically related purposes: (1) identifying areas for improvement in individual programs, and (2) rating and ranking of individual programs for the purpose of competitively awarding merit and bonus funds (as stipulated by Congress in the 2002 legislation, P.L.107–299). These purposes are related insofar as competition for funds serves as an incentive to the individual programs to improve. The evaluation process as it has been performed since 2003, however, appears to be more appropriately structured to achieve the narrow goal of ranking programs and distributing competitive funds. For the overall program to improve—and, in particular, for it to become (and be seen to become) a truly national program—there is need for NSGO to strengthen its ability to facilitate and coordinate efforts of the individual programs.

Perhaps the foremost concern about the Sea Grant evaluation process is the reliance by the NSGO on periodic assessments as the primary, if not only, means of evaluation and oversight. Despite the general high quality of the information they provide, the overreliance on periodic assessments undermines the role that the NSGO could play in continued improvement of the individual programs and in the administration and coordination of the national program. The periodic assessments themselves rely heavily on information collected during quadrennial visits by Program Assessment Teams (PATs) overseen by the NSGRP. As the members of PATs and the NSGRP are not Sea Grant employees, the preponderance of program oversight is actually external. **The Director of the National Sea Grant College Program, under supervision of the Secretary of Commerce and in consultation with the National Sea Grant Review Panel and the individual Sea Grant programs, should strengthen the ability of the National Sea Grant Office to carry out meaningful, ongoing internal assessment in order to complement periodic, external assessment currently taking place.**

STRATEGIC PLANNING

The importance of strategic planning in program development, implementation, and evaluation was emphasized in the 1994 NRC report. Specifically, the 1994 report recommended that "State Sea Grant Directors [individual Sea Grant Program directors] and the Director of the NSGCP [National Director] must cooperate to develop a single strategic plan articulating a shared vision and strategies which must be fully integrated into, and reflective of, NOAA's strategic plan." Although strategic planning at the national level, as carried out by the NSGO, meets this recommendation, the same cannot be said at the state level. More effort is needed to ensure that all of the individual Sea Grant programs develop strategic plans that dovetail with the national plan, while addressing local and state challenges they may be uniquely equipped to address.

Since 1994, a number of high-level reviews, such as the recent report from the U.S. Commission on Ocean Policy, have identified the most pressing problems in marine policy. These reviews have repeatedly emphasized the need to identify and address issues at the proper regional scale. The state and federal partnership NSGCP represents would seem to be well suited to addressing these intermediate-scale problems, as federal coordination and support for local and state efforts is generally an important component to effective regional action. To ensure that strategic planning reflects a shared vision, representatives of the NSGO should participate in the local strategic planning process and the strategic plan should serve as the basis upon which the individual Sea Grant program is evaluated. **Steps should be taken by the Director of the National Sea Grant College Program, under supervision of the Secretary of Commerce and in consultation with the National Sea Grant Review Panel and the individual Sea Grant programs, to strengthen strategic planning at both the**

national and individual program level. The strategic plans of the individual programs and the national program should represent a coordinated and collective effort to serve local, regional, and national needs. As discussed in Chapter 4, actions by the NSGO should include:

- developing and implementing a process to assist individual programs in strategic planning; and
- creating a separate process for evaluating and approving appropriately ambitious strategic plans for the individual programs.

PERFORMANCE CRITERIA

Performance criteria are a combination of quantitative and qualitative measures used to assess the selected activity or program, the outcomes of that program, and, in some instances, the system that program is intended to influence. In the case of assessing the effectiveness and impacts of individual Sea Grant programs, this involves assigning benchmarks to describe the expected level of performance in a particular area (such as program organization and management) and indicators to help assess the outcomes or impacts of the individual program in that area. As discussed earlier, strategic planning is a critical basis for implementation, review, and evaluation of institutional programs. **The Director of the National Sea Grant College Program, under supervision of the Secretary of Commerce and in consultation with National Sea Grant Review Panel and the individual Sea Grant programs, should modify the benchmarks and indicators, as needed, to ensure that the performance of each program is measured against the objectives outlined in the separately approved, program specific strategic plan called for in the previous recommendation.**

In addition, the current Sea Grant evaluation criteria do not recognize the importance individual programs should play in building cooperative efforts to address regional and even national scale problems. The existing benchmarks tend to encourage program development at the local scale. Furthermore, the use of the periodic assessment scores in determining merit and bonus allocations may have resulted in lower levels of cooperative behavior between programs, which now see themselves as pitted against one another. Encouraging programs to undertake cooperative efforts to address regional scale problems thus needs to be incorporated into Sea Grant evaluation criteria and given a high value.

Modifying the evaluation criteria to place greater weight on cooperative efforts is not intended as a recommendation to increase the complexity of the criteria. To the contrary, the current set of scored criteria are found to be overly complex and numerous, requiring significant amounts of time be devoted to developing consensus scores on a large number of criteria that, in many cases, account for a small percentage of the overall score. This endeavor to achieve greater precision by increasing the number of score criteria tends to inadvertently discourage efforts to produce more holistic judgments of program performance. **The Director of the National Sea Grant College Program, under supervision of the Secretary of Commerce and in consultation with National Sea Grant Review Panel and the individual Sea Grant programs, should substantially reduce the overall number of scored criteria by combining various existing criteria, while adding cooperative, network-building activities as an explicitly evaluated, highly valued criterion.** Implementation of revised criteria should be postponed until the beginning of the next cycle of program review (the current review cycle will conclude in late 2006).

PROGRAM ASSESSMENT TEAMS AND SITE VISITS

Two of the major shortcomings of the current program assessment process are the limited overlap of the PATs in membership and the inability to evaluate the entire program in less than four years. Together, these problems compromise the reliability and credibility of the annual ranking required under the 2002 Act Amendments. Both shortcomings could be alleviated to a degree by shortening the PAT site visits and focusing attention during the visit on the most essential evaluation tasks. Reducing the demands on the PATs would allow members to serve on more than one team and would also allow a larger number of site visits each year. As long as the PAT process remains the primary source of information to rate and rank individual programs, steps will need to be taken to improve the reliability and credibility of the process. **The Director of the National Sea Grant College Program, under supervision of the Secretary of Commerce and in consultation with the National Sea Grant Review Panel and the individual Sea Grant programs, should shorten the duration of and standardize the PAT site visits, based on the minimum time and material needed to cover essential, standardized elements of the program assessment.** If, as recommended in chapters 3, 4, and 5 of this report, the annual evaluation process carried out by the NSGO is modified so as to provide a reliable and credible assessment of individual programs, changes to the PAT process to improve reliability will be less urgent. This would allow greater flexibility for the scope and design of PAT visits.

PROVIDING COORDINATION AND FACILITATION THROUGH INFORMED, ONGOING OVERSIGHT

Greater involvement and ongoing oversight by the NSGO is needed to ensure that the program as a whole continues to improve while addressing, local, regional and national needs. Informed oversight is also needed to lend credibility to annual program rankings and the allocation of merit and bonus funds. The two goals of program improvement and increased credibility can be simultaneously served by a meaningful ongoing, annual evaluation process that complements the periodic assessment carried out during the PAT site visit. This annual evaluation process, discussed in detail in Chapter 4, would replace the current NSGO Final Evaluation Review (FE). The current FE is summarized in Chapter 2. Review material prepared for the annual review should include an effective annual report, supplemented by material that demonstrates the extent to which the annual activities combine to form a cohesive ongoing program of activity organized to accomplish the objectives of appropriately ambitious strategic plans. **The Director of the National Sea Grant College Program, under supervision of the Secretary of Commerce, should rank the individual Sea Grant programs based on a program evaluation process that includes more robust, credible, and transparent annual assessments of each individual Sea Grant program.** Assessment of programs that have undergone periodic assessments in the preceding year should also include consideration of the PAT reports and the individual Sea Grant program directors' responses to the PAT reports. The additional effort required of individual Sea Grant programs to provide information on an annual basis can be offset to a degree by reducing the time required to prepare materials for the periodic assessment, if most of the information required by the latter can be made up of materials submitted annually.

FAIRNESS IN COMPETITION

This study systematically evaluated the possibility that assessments, ratings, and the subsequent ranking of program performance are influenced by size or age of the program, location, type of institutional administration linkages, and years of experience of the program officer within the NSGO. With one exception, a statistical analysis relating program ratings with these and other factors found no significant bias. The exception is a positive correlation between years of experience of the program officer with the program under evaluation and the improvement in program score during the FE. Although the changes in program score were generally fairly small, the nature of the current assessment and ranking process results in a very narrow range of program scores overall; thus, even a minute difference in the score assigned to two similarly performing programs that straddle the boundary between bonus categories could result in a significant difference in the amount of bonus funds awarded. **The Director of the National Sea Grant College Program, under supervision of the Secretary of Commerce, should revise the calculation of bonus funding allocation relative to program rank to ensure that small differences in program rank do not result in large differences in bonus funding, while preserving or even enhancing the ability to competitively award bonus funds as required by the National Sea Grant College Program Act Amendments of 2002 (P.L. 107–299).**

IMPROVING PROGRAM COHESION

The NSGO does not currently play a sufficient role in ongoing program assistance, communication, and assessment, or in maintaining close ongoing working relationships with the individual Sea Grant programs. This limits the ability of the NSGO, and by extension the National Director, to "provide an appropriately balanced response to local, regional, and national needs, which is reflective of integration with the relevant portions of strategic plans of the Department of Commerce and of the Administration" (33 U.S.C. 1123).

In order for the NSGO to more effectively administer the NSGCP and coordinate and facilitate the efforts of the individual Sea Grant programs, thus fulfilling the federal role within the Sea Grant partnership, the capabilities of the NSGO should be reevaluated, and likely, enhanced. **The Secretary of Commerce, in consultation with the National Sea Grant Review Panel, should take steps to ensure that sufficient human and fiscal resources are available to allow robust, ongoing, and meaningful interaction among the Director of the National Sea Grant College Program, the staff of the National Sea Grant Office, the directors of individual Sea Grant programs, and the administrators of the home institutions of individual Sea Grant programs.**

While the purpose of this study was not to provide specific recommendations about how the NSGO should be organized, staffed, or funded, it does seem appropriate to point out various approaches that might be considered for achieving this recommended action, without significantly expanding the size of the NSGO staff. One such approach might include establishing a small number of program officers who spend a far greater portion of their time working with a small number of individual programs with common challenges than is currently possible now. Indeed, additional approaches need to be further explored. **The Director of the National Sea Grant College Program, under supervision of the Secretary of Commerce and in consultation with the National Sea Grant Review Panel and the individual Sea Grant programs, should undertake an evaluation of how work force capabilities and other compo-**

nents of effective program administration could be modified within the National Sea Grant Office to enhance its ability to coordinate and facilitate the actions of individual Sea Grant programs.

Based on comments made during information gathering meetings, written correspondence submitted in response to committee requests, and various NSGO and NSGRP documents, it is apparent that a number of individual program directors remain confused about key aspects of the program assessment process, the annual evaluation process, and their impacts on program rankings and funding. Although responsibility for understanding this process rests with the individual Sea Grant program directors, the NSGO has a responsibility to make sure the process is reasonably straightforward and understandable. As discussed in Chapter 3, there should be greater attention and clarity regarding all aspects of program assessment. **The Director of the National Sea Grant College Program, under supervision of the Secretary of Commerce, should take steps to ensure that the program assessment process (both the new annual assessment called for in this report and the PAT review) is well described and understood by individual program directors, congressional staff, personnel of the Office of Management and Budget, university and state administrators, and the general public.**

If the recommendations put forth above are implemented, the functions of the annual and periodic assessments will evolve such that both will provide different and independent sources of information about the state of the Sea Grant program as a whole. This information should provide important insights to the Secretary of Commerce, the National Director, and potentially Congress. Thus, there would seem to be a need to synthesize and analyze the results of these assessments every four years, including a synthesis of all periodic assessments completed during that time and a systematic review of the NSGO. Developing such a "state of the Sea Grant program" report would seem to be an obvious role for the NSGRP. **The Director of the National Sea Grant College Program, acting under authority of the Secretary, should direct the National Sea Grant Review Panel to undertake the development of a systematic review of the "state of the Sea Grant program" once every four years. The review should rely extensively on information collected during the annual and periodic assessments, augmented with a site visit to the National Sea Grant Office, and it should focus on how the program is functioning as a whole.** In addition to commenting on how the program is performing in terms of the various criteria used during the assessments, the "state of the Sea Grant program" report could identify needed changes in how the program is administered, how the assessment process is carried out, or other areas as deemed valuable by the Secretary of Commerce or the National Director.

1

Introduction

The National Sea Grant College Program (NSGCP) is a network of 30[1] individual Sea Grant programs[2] and provides funds via these programs for marine and Great Lakes applied research, education, and outreach. Sea Grant has been a major source of funding in the United States for work in areas such as marine aquaculture, shellfish disease, aquatic nuisance species, coastal and estuarine ecology, seafood safety, marine biotechnology, marine engineering, marine technology development, and marine policy. Each of the 30 individual Sea Grant programs (see Figure 1.1) facilitates communication among university researchers,

---

[1]Of the 30 Sea Grant programs, 28 are individual Sea Grant College programs and 2 are Sea Grant Institutional programs. California and Massachusetts have two Sea Grant programs each, namely the University of California (located at Scripps Institution of Oceanography) and the University of Southern California (USC) programs and the Massachusetts Institute of Technology and the Woods Hole Oceanographic Institution (WHOI) programs, respectively. USC and WHOI are the two Sea Grant Institutional programs. Two individual Sea Grant programs operate as bi-state programs (Mississippi-Alabama, Illinois-Indiana). In addition to these 30 Sea Grant programs, Pennsylvania, Vermont (Lake Champlain), and Guam (3 programs) are in the initial stages of developing a full Sea Grant program and have not yet been included in the evaluation process. Source: F. Schuler, NOAA, personal communication, 2005.

[2]For the purpose of this report, each of the 30 programs will be referred to as an "individual Sea Grant program" to differentiate it from the entire network, which is referred to as the National Program or NSGCP. Previous Sea Grant literature and legislation have also used the terms "state program" and "Sea Grant college/institute."

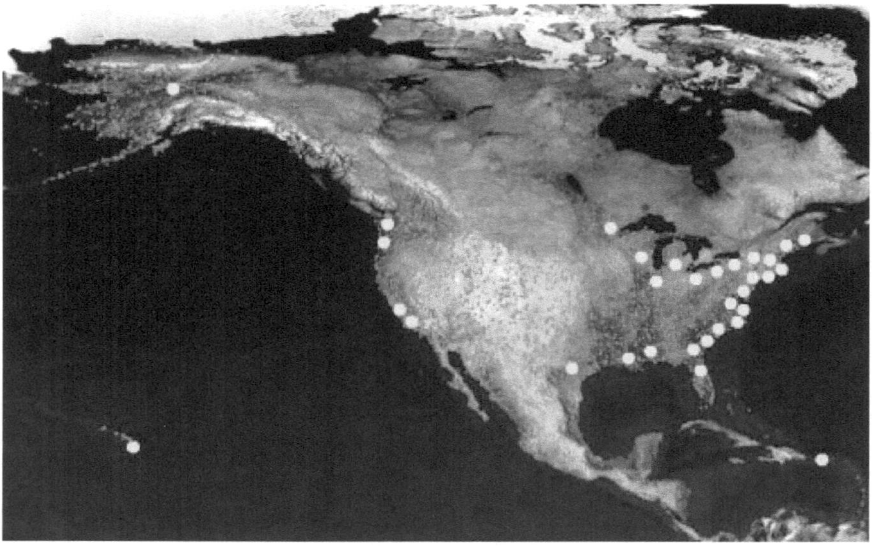

FIGURE 1.1 National Sea Grant College Program Network (Guam outside of map range). Source: National Oceanic and Atmospheric Administration (NOAA).

industry members, policy makers, educators, and the public. Through its outreach and extension services, scientific research results are shared with the user communities, and these groups in turn communicate their problems and needs back to the researchers. Thus, Sea Grant plays an important role in identifying problems, funding potential solutions, and providing educational opportunities and materials. There are Web sites[3] where individual Sea Grant program directors and the general public can obtain information on the NSGCP and all funded projects.

ORIGIN OF THE NATIONAL SEA GRANT COLLEGE PROGRAM

The idea of a Sea Grant college program was originally put forward by oceanographer, inventor, and writer Dr. Athelstan Spilhaus at the 93rd meeting of the American Fisheries Society in 1963. Interest in the Sea

[3]The National Sea Grant Office website (http://www.seagrant.noaa.gov), the National Sea Grant Library (http://nsgd.gso.uri.edu/), the National Sea Grant Law Center (http://www.olemiss.edu/orgs/SGLC/lawcenterhome.htm), the National Sea Grant Education Teacher Resource—The Bridge (http://www.vims.edu/bridge/), and the Sea Grant Media Center (http://www.seagrantnews.org/).

Grant concept grew, much of it sparked by an editorial written by Spilhaus (1964, p. 993):

> . . . . I have suggested the establishment of "sea-grant colleges" in existing universities that wish to develop oceanic work. . . . These would be modernized parallels of the great developments in agriculture and the mechanic arts which were occasioned by the Land Grant Act of about a hundred years ago. . . . Establishment of the land-grant colleges was one of the best investments this nation ever made. The same kind of imagination and foresight should be applied to exploitation of the sea.

In 1965, U.S. Senator Claiborne Pell of Rhode Island introduced legislation to establish Sea Grant colleges on campuses nationwide as centers of excellence in marine and coastal studies. With the adoption in 1966 of the National Sea Grant College and Program Act (P.L. 89–688) (see Appendix C for key Sea Grant program legislation), Congress established a federal government/academic/industry partnership supporting the "establishment, development, and operation of programs by sea grant colleges and . . . other sea grant programs designed to achieve the gainful use of marine resources" (P.L. 89–688). The development of marine resources was defined as:

> . . . scientific endeavors relating to the marine environment, including but not limited to the fields oriented toward the development, conservation, or economic utilization of the physical, chemical, geological and biological resources of the marine environment, the fields of marine commerce and marine engineering, the fields relating to exploration or research in, the recovery of natural resources from, and the transmission of energy in, the marine environment; the fields of oceanography and oceanology and the fields with respect to the study of the economic, legal, medical or sociological problems arising out of the management, use, development recovery and control of the natural resources of the marine environment [P.L. 89–688].

The term marine environment was defined in the Act as: "the oceans, the Continental Shelf of the United States, the Great Lakes, the seabed and subsoil of the submarine areas adjacent to the coasts of the United States to the depth of two hundred meters or beyond that limit" (P.L. 89–688).

Title 33, Chapter 22 of U.S. Code,[4] The National Sea Grant College Program Act (33 U.S.C. 1121–1131) codified P.L. 89–688, and subsequent

[4] The U.S. Code is the codification by subject matter of the general and permanent laws of the United States. It is divided by broad subjects into 50 titles and published by the Office of the Law Revision Counsel of the U.S. House of Representatives.

amendments (e.g., P.L. 94–461,[5] P.L. 105–160, P.L.107–299) (see Appendix H for reprinting of most sections in Chapter 22). In Section 1121 of Title 33, Congress declares the following policy:

> . . . . The understanding, assessment, development, utilization, and conservation of such resources require a broad commitment and an intense involvement on the part of the federal government in continuing partnership with State and local governments, private industry, universities, organizations, and individuals concerned with or affected by ocean, coastal, and Great Lakes resources. . . . The National Oceanic and Atmospheric Administration, through the national sea grant college program, offers the most suitable locus and means for such commitment and involvement through the promotion of activities that will result in greater such understanding, assessment, development, utilization, and conservation. The most cost-effective way to promote such activities is through continued and increased federal support of the establishment, development, and operation of programs and projects by sea grant colleges, sea grant institutes, and other institutions, including strong collaborations between administration scientists and scientists at academic institutions [33 U.S.C. 1121].

U.S. CODE: LEADERSHIP ROLES AND RESPONSIBILITIES DEFINED

Title 33, Chapter 22 of the U.S. Code defines the responsibilities of the key components of the National Sea Grant College Program, and those of various other entities within the federal government. This section defines those responsibilities.

The Secretary of Commerce, acting through the Under Secretary of Commerce for Oceans and Atmosphere,[6] is responsible for maintaining the National Sea Grant College Program (referred to as "NSGCP" or "National Program" throughout this report[7]), which is to be administered by

[5]P.L. 94–461 completely rewrote the Congressional statement of findings, objectives, and purposes of the National Sea Grant Program Act to reflect the extension and strengthening of the national sea grant program to promote research, education, training, and advisory service activities in fields related to ocean and coastal resources through federal support to sea grant colleges, sea grant regional consortia, and other institutions through NOAA, and to make education, training, research, and advisory services responsive to state, local, regional, or national needs and problems.

[6]The term "Secretary" refers to the "Secretary of Commerce, acting through the Under Secretary of Commerce for Oceans and Atmosphere" (33 U.S.C. 1122 [15]). Currently, VADM Conrad Lautenbacher is the Under Secretary of Commerce for Oceans and Atmosphere as well as the Administrator of NOAA.

[7]33 U.S.C. 1122 uses "The Program" to refer to the National Sea Grant College Program.

the National Sea Grant Office (NSGO). To carry out this function, the Secretary appoints the Director of the National Sea Grant College Program ("National Director"), who, subject to the supervision of the Secretary, administers the NSGCP and oversees the operation of the NSGO. Thus, the Secretary is ultimately responsible for the appointment, assignment of duties, transfer, and compensation of "such personnel as may be necessary, to administer the Program." In addition, the Secretary, in consultation with the National Sea Grant Review Panel (NSGRP) and the individual Sea Grant programs (both discussed below), and acting through the National Director, establishes guidelines related to the activities and responsibilities of the individual Sea Grant programs. These guidelines are the major input into the development, every four years, of a strategic plan that establishes priorities for the National Program, provides an appropriately balanced response to local, regional, and national needs, and is reflective of integration with relevant portions of the strategic plans of the Department of Commerce and the Administration (NOAA[8]; 33 U.S.C. 1123).

The National Director is appointed by the Secretary of Commerce to administer the NSGCP and oversee the operation of the NSGO. The National Director, subject to the supervision of the Secretary and in consultation with the NSGRP and individual Sea Grant programs, facilitates and coordinates the development of a strategic plan every four years that establishes priorities for the National Program (33 U.S.C. Sec. 1123). In addition, the National Director encourages the establishment and growth of individual Sea Grant programs and facilitates the cooperation and coordination of the National Program with other Federal activities in fields related to ocean, coastal, and Great Lakes resources. The National Director is also charged with evaluating the performance of the individual Sea Grant programs and rating the programs according to their relative performance. Title 33 U.S.C. 1123 (d)(3)(A) prescribes that the National Director rank the individual Sea Grant programs "into no less than 5 categories, with each of the 2 best-performing categories containing no more than 25 percent of the programs." Title 33 U.S.C. 1123 (d)(3)(B) prescribes that the National Director, subject to the availability of appropriations, allocate funding among individual Sea Grant programs so as to: (i) promote healthy competition among the individual Sea Grant programs; (ii) encourage successful implementation of the individual programs; (iii) to the maximum extent consistent with other provisions of *The National Sea Grant College Program Act* provide a stable base of funding for individual

[8]The term "Administration" refers to the National Oceanic and Atmospheric Administration (Title 33 U.S.C. 1122 [1]).

Sea Grant programs; and (iv) encourage and promote coordination and cooperation among the research, education, and outreach programs of NOAA and those of academic institutions (individual programs).

The Directors of individual Sea Grant Colleges and Institutes (referred to as directors of individual Sea Grant programs in this report) are required by Title 33 U.S.C. 1126 to coordinate program activities and help set local, regional and national priorities. Thus the directors of the 30 Sea Grant programs (see Figure 1.1 for map of current Sea Grant locations) play a prominent and pivotal role in carrying out the function of the National Program. In addition to overseeing the merit review of all proposals for grants and contracts awarded under authority provided by Title 33 U.S.C. 1124, it is the responsibility of each individual director, in consultation with the National Director and the National Sea Grant Review Panel, to develop and implement a program that is consistent with the guidelines and priorities established by the National Strategic Plan required by Title 33 U.S.C. 1123 (c)(1). Furthermore, each individual Sea Grant program administers a significant pool of nonfederal funds, provided either as a match to federal funding, or as a grant or contract with a state or local funding source. When acting collectively through the Sea Grant Association (SGA) (to be discussed shortly), the directors of the individual Sea Grant programs are a unified voice for these institutions on issues of importance to the oceans and coasts.

NOAA's National Sea Grant Office (NSGO), as mandated by Title 33 U.S.C. 1123 (a), operates under the direction of the National Director and administers funding to the individual Sea Grant programs and oversees several national funding competitions. The NSGO also facilitates the Department of Commerce designation of Sea Grant College programs[9] and oversees the program assessment process. The NSGO, in consultation with the NSGRP and individual Sea Grant programs, is responsible for the development of a strategic plan that establishes priorities for the NSGCP, provides an appropriately balanced response to local, regional, and national needs, and is reflective of integration with relevant portions of the strategic plans of the Department of Commerce and NOAA. In addition, the NSGO is responsible for managing funding competitions for National Strategic Investments; three fellowship programs (i.e., the John A. Knauss Marine Policy Fellowship, the Sea Grant/NOAA Fisheries

[9]Designation of an individual Sea Grant program is the official naming of an institution of higher education or confederation of such institutions as an official Sea Grant College program as bestowed by the Secretary of Commerce. Applicant institutions must meet certain eligibility requirements.

Graduate Fellowship, the Sea Grant Industry Fellowship Program); and providing national coordination and leadership for Sea Grant's research, education, extension, communications, and fiscal networks. By law, the NSGO must use no more than 5 percent of the total budget for administrative costs in any given fiscal year to administer the NSGCP (33 U.S.C. 1131).

The National Sea Grant Review Panel (NSGRP), as mandated by Title 33 U.S.C. 1128, comprises 15 individuals with diverse backgrounds in marine affairs. The panel, appointed by the Secretary of Commerce, is charged with advising the Secretary and the National Director concerning: (i) applications or proposals for, and performance under, grants and contracts awarded; (ii) the Sea Grant Fellowship Program; (iii) the designation and operation of Sea Grant Colleges and Institutes, and the operation of Sea Grant Programs; (iv) the formulation and application of the planning guidelines and priorities (as discussed above); and (v) "such other matters as the Secretary refers to the panel for review and advice"(33 U.S.C. 1128).

In 1998, in partial response to the 1994 National Research Council (NRC) report *A Review of NOAA National Sea Grant College Program* and the 1997 *Report on Evaluation of Sea Grant College Programs* requested by the NSGO and completed by Copeland et al. (1997), the National Director, acting under the supervision of the Secretary of Commerce, exercised authority under Title 33 U.S.C. 1128(b)(5) to request that the NSGRP formally oversee the periodic assessment of individual Sea Grant programs (required by Title 33 U.S.C. 1123[c][2] as amended by the National Sea Grant College Program Reauthorization Act of 1998 [P.L. 105–160]).

The Sea Grant Association (SGA) is a nonprofit organization comprising the academic institutions that participate in the NSGCP (i.e., primarily directors and other administrators of individual Sea Grant programs). Though not a formal part of the NSGCP, the SGA plays an important role in furthering the Sea Grant program concept. The SGA provides the mechanism for these institutions to coordinate the research, education, training, and outreach activities of individual Sea Grant programs and to set program priorities (to enhance the economic, environmental, and social potential of the nation's coastal, marine, and Great Lakes resources) at both the regional and national level (SGA Brochure, available online at http://www.sga.seagrant.org).

THE EVOLVING SEA GRANT PROGRAM REVIEW PROCESS AND ENABLING LEGISLATION

This report is the product of the NRC study requested by Congress in P.L.107–299, sponsored by the Under Secretary of Commerce for Oceans and Atmosphere, and completed in 2006. It is the second NRC review of the Sea Grant program.

The first NRC study, requested by the Under Secretary of Commerce for Oceans and Atmosphere in 1993 and completed in 1994, reviewed and evaluated the NSGCP to provide the basis for any needed changes to the program and to provide information for NOAA as it worked with Congress on the then pending National Sea Grant College Program Reauthorization Act of 1998. The statement of task for the first study focused on the entire program, and the resulting report, *A Review of the NOAA National Sea Grant College Program*, made several recommendations for improving the program overall (Box 1.1).

With regard to the proposal and program review process, the 1994 NRC report suggested that the review process for research proposals be decoupled from the NSGO evaluation of individual Sea Grant programs. It also recommended that standard scientific and peer review procedures be implemented for all of the individual Sea Grant programs. The report recommended that the review process and all aspects of program implementation, including administration, be streamlined prior to FY 1996. In addition, the report called for the NSGO to evaluate the success of each individual program on a four-year cycle, using, in part, retrospective information on recent achievements, based on measures for each of the three areas of research, education, and outreach. Finally, it was recommended that the NSGRP evaluate the performance of the NSGO on the same timetable.

Following the release of the first NRC study and other efforts by Congress, NOAA, and other key players, the National Sea Grant College and Program Act was reauthorized by Congress in both 1998 and 2002 (P.L. 105–160 and P.L. 107–299, respectively).

Some Highlights of the National Sea Grant College Program Legislation

In the 1998 reauthorization of the National Sea Grant College and Program Act, Congress made some changes to the NSGCP. Among the more notable changes was the establishment of the performance based evaluation system, or PAT review, and the direction that funded resources would be allocated to programs based, in part, on their performance.

Congress enacted several new program requirements in the 2002 reauthorization of the National Sea Grant College Program. Three of these new requirements are relevant to this study (P.L. 107–299):

Box 1.1
Key Issue Areas and Recommendations from A Review of the NOAA National Sea Grant College Program (Reprinted from NRC, 1994)

ISSUE 1—SEA GRANT'S POSITION WITHIN NOAA
The Administrator must ensure that NSGCP has *appropriate* responsibility and capability for research, education and outreach across NOAA. NSGCP should be relocated within NOAA to report directly to the Office of the Administrator.

ISSUE 2—SHARED VISION AND STRATEGIC PLANNING
State Sea Grant directors [individual Sea Grant program directors] and the Director of the NSGO [National Director] must cooperate to develop a single strategic plan articulating a shared vision and strategies which must be fully integrated into, and reflective of, NOAA's strategic plan. Unified Sea Grant strategic planning should begin immediately so that its results can be incorporated in the Fiscal Year (FY) 1997 NOAA budget.

ISSUE 3—OVERLAPPING ROLES AND RESPONSIBILITES
The roles and responsibilities of the state Sea Grant directors [individual Sea Grant program directors], NSGO, and National Sea Grant Review Panel [NSGRP] must be clarified. The resultant roles and responsibilities of NSGO and NSGRP should be clarified by the NOAA Administrator prior to the 1995 reauthorization.

ISSUE 4—PROPOSAL REVIEW AND PROGRAM EVALUATION
The review process for research proposals should be decoupled from the NSGO evaluation of state programs prior to the 1995 reauthorization. Standard scientific and peer review procedures should be implemented for all state Sea Grant programs. The review process and all aspects of program implementation, including administration, should be streamlined prior to FY 1996. NSGO should evaluate the success of each state program on a four-year cycle, using, in part, retrospective information on recent achievements, based on measures for each of the three areas of research, education, and outreach. NSGRP should evaluate the performance of NSGO on the same timetable.

ISSUE 5—INTERACTIONS WITH INDUSTRY
NSGO and the state Sea Grant Programs must increase their interactions with marine industry to include program policy guidance, expanded outreach and marine advisory services, joint research projects, and substantial industry financial support of the Sea Grant program. Action to address this recommendation should form part of the examination of the performance of each state program. These actions should be identified in the Sea Grant strategic plan.

ISSUE 6—FUNDING
The committee agreed that NSGCP needs additional funding to fulfill its potential. In the last decade, the purchasing power of the average research grant has declined by about one-half. A steady increase in funding is necessary if the program's potential contributions to the nation's economic and environmental health are to be realized. Any additional funds appropriated to NSGCP should be split between enhancement of meritorious state programs and support of new initiatives.

- **Strategic Planning**: The Secretary of Commerce was directed to develop a strategic plan every four years and to consult and coordinate with the NSGRP and individual Sea Grant programs when doing so.
- **New Rating and Ranking of Sea Grant Programs**: The Secretary, acting through the National Director, was directed to evaluate the performance of individual Sea Grant programs and rate these programs using the priorities, guidelines, and qualifications established by the Secretary, and rank the programs according to their relative performance into no less than 5 categories and with each of the 2 best-performing categories containing no more than 25 percent of the programs.
- **Review the Evaluation Process**: A review of the Sea Grant evaluation and rating process was requested by the Act. The National Academies was to start this review three years from the date of enactment (enactment was November 26, 2002). The U.S. Department of Commerce was directed to have the National Academies review the effectiveness of the evaluation and rating system (under the 2002 amendment) in determining the relative performance of programs of individual Sea Grant programs, and to evaluate whether the individual Sea Grant programs have improved as a result of the evaluation process. The National Academies was also requested to make recommendations to improve the overall effectiveness of the evaluation process.

STUDY APPROACH AND REPORT ORGANIZATION

In response to congressional mandate (P.L. 107–299), the National Academies formed a committee of experts to carry out evaluate the NSGCP review process (see Box 1.2 for specific statement of task).

Study Approach

Information to support the study's conclusions was gathered through direct requests and public meetings. Materials and comments were requested from the NSGO, the NSGRP, the SGA, and from all individual Sea Grant program directors. Three public meetings were held in: Washington, D.C. (March 2–4, 2005); Rockport, Maine (June 4–5, 2005), concurrently with the first few days of the biennial Sea Grant Week; and Ann Arbor/Detroit, Michigan (August 9–11, 2005). During those meetings, the committee heard presentations by staff of the SGA and the NSGO. Open forum sessions were held where directors of individual Sea Grant programs shared concerns and observations (see Box 1.3 for some questions asked by the Committee at an open forum). The committee also had one-on-one discussions with several individual Sea Grant program directors

Box 1.2
Statement of Task

This study will assess new procedures adopted by the National Sea Grant Program since the publication of the 1994 National Research Council report *A Review of NOAA National Sea Grant College Program* to determine their impacts. During this study, the committee will address the impact of the new procedures and evaluation process on Sea Grant as a whole, identifying constructive changes and value added to overall institutional effectiveness, responsiveness, quality of management, leadership, and reputation.

As part of this assessment, the committee will examine:

(1) Effectiveness of major changes instituted in response to the recommendations of the 1994 NRC report with regard to individual program performance and quality.

(2) Effectiveness of program review procedures with regard to accuracy, accountability, and enhancement of individual program performance. Both the previous and current (adopted in 2003 in response to the Sea Grant Act of 2002 [P.L. 107-299]) review procedures will be assessed as specified below:

- Review the effectiveness of the evaluation and rating system in determining relative performance of programs with regard to management and quality of research, education, extension, and training activities;
- Evaluate whether there have been improvements in programs as a result of the evaluation process;
- Evaluate the 2003 review procedures for their ability to meaningfully segregate individual programs into five categories based on competitive scores; and
- Compare the effectiveness of the previous and 2003 review procedures with regard to the dual objectives of maximizing the quality of each program and of rating programs relative to each other for the purpose of determining performance-based funding.

(3) Assessment of the usefulness and fairness of metrics developed to evaluate programs with different operational constraints, resources, and local priorities.

- Evaluate metrics for relevance and clarity;
- Determine whether metrics provide a quantitative measure of quality of performance; and
- Assess whether metrics improve consistency and objectivity of reviews from different teams evaluating a diverse portfolio of state Sea Grant programs.

The committee will make recommendations for improving the overall effectiveness of the evaluation process to ensure fairness, consistency, and enhancement of performance.

Box 1.3
Questions Asked of all Attendees at the NRC Committee Forum on June 5, 2005, in Rockport, ME

IMPROVEMENTS
Has the effectiveness of review been improved by the changes since 2002? Apart from your program, are the best programs receiving the best scores?

EFFICIENCY OF EVALUATIONS
Does the expense and effort justify the outcome? If Sea Grant programs need to be ranked would you prefer to use the PAT process or do you have other suggestions?

STANDARDIZATION OF PAT REVIEWS/FINAL EVALUATION
Do benchmarks adequately capture program outcomes . . . for education? . . . for extension? . . . for outreach? . . . for research?
Are we measuring what we care about or caring about what we measure?
Is the use of weights for the subcategories appropriate? Do NSGO staff and program officers use the same weights and benchmarks consistently throughout time and for each program? Are PAT manuals and benchmarks shared throughout your program?
Is there adequate consistency of PAT teams between reviews (to individual SG directors that have served on less than one PAT)?
Have you expressed interest in serving on a PAT and not been invited?

SUGGESTIONS
What is the primary change you would make to the program assessment? Why?

ROLE OF PROGRAM OFFICER
What do you think the role of the program officer should be in general?

REGIONAL COOPERATION AND COLLABORATION
Have there been recent efforts to stimulate collaboration between programs? Do individual directors think collaboration a good idea? What's hindering the process? Is it valued in the evaluation process?

and their staff members regarding the evaluation process and its impact on the individual programs.

During the two-year study, the Committee observed a number of Program Assessment Team visits to individual Sea Grant programs—several day meetings where individual Sea Grant programs are reviewed by an assigned Program Assessment Team (detailed discussion in Chapter 2)—in the states of Washington (2004), Oregon (2005), Georgia (2005), Ohio (2005), New York (2005), and Massachusetts (2005, Woods Hole Oceanographic Institution program). Two representatives of the committee attended the last few days of the week-long nonpublic meeting of the NSGO (February 2005), referred to throughout this report as "NSGO Final Evaluation Review." In addition to these efforts, NRC staff met and corresponded with U.S. Office of Management and Budget staff, congressional staff, and NSGO staff to obtain specific information and historical data.

In addition, the committee contacted each individual Sea Grant program and requested information on previous and current PATs (Cycle 1 and Cycle 2), PAT reports, director response letters, final evaluation letters, information on costs of preparing for and conducting PAT reviews during Cycle 2, PAT briefing materials, and answers by individual Sea Grant directors to the questions raised in a formal letter written by the chair of the Committee (see Appendix I for the text of this letter sent to each individual Sea Grant program director). Almost all (28 out of the 30 reviewed Sea Grant programs) Sea Grant programs responded to this request. Individual programs sent many of the materials from both Cycle 1 and Cycle 2 reviews, including: PAT reports, director's responses to the final evaluation letter; funding allocation letters, PAT briefing materials, and information specifically on Cycle 2 costs incurred (as requested by the Committee). In addition, individual Sea Grant program directors submitted letters and comments, information on the Topical Advisory Teams (TATs), and miscellaneous additional documents.

Further, the committee reviewed all key documents written on the Sea Grant review process to date (Byrne et al., 2000; Toll et al., 2001; Duce et al., 2002; Kudrna et al., 2005;[10]see reference lists of these reports), the last eight years of PAT manuals (from 1998 to 2005), and many of the documents provided on the NSGO Sea Grant shared database. The findings and recommendations of the committee were based on all of this research and their own experience.

The Structure of the Report

This report attempts to identify strengths and weaknesses of the current evaluation process and suggests improvements to enhance the performance of the individual programs and the Sea Grant program as a whole. Chapter 2 discusses the history of the Sea Grant program review process, thereby providing context for subsequent analyses. Chapter 3 explains and critiques the current assessment process (both the PAT review and NSGO Final Evaluation Review) and provides recommenda-

[10]Executive summary of the Kudrna et al. (2005) is provided in Appendix J.

tions for improving the reliability, efficiency, and transparency of the competitive process. Chapter 4 discusses the broader need for program oversight and management and makes suggestions for how to move beyond the periodic assessment process in an effort to strengthen NSGCP efforts to provide an appropriately balanced response to local, regional, and national needs. Finally, Chapter 5 discusses report findings and recommendations as a whole, summarizing key findings and recommendations from chapters 3 and 4 in an integrated narrative.

2

History of Sea Grant Program Review and Assessment

The National Sea Grant College Program's (NSGCP) processes for overseeing and evaluating the individual Sea Grant programs have evolved since the program's inception. One of the more notable changes in the process has been the increasing use of peer review and the administrative level at which it is carried out. Another notable change has been the emergence of external periodic assessment as a tool to support the distribution of merit and bonus funding, rather than simply to identify areas for program improvement. These changes have affected the make-up and role of National Sea Grant Office (NSGO). Prior to 1994, the NSGO was organized around program officers and specialists assigned to monitor institutional programs and exercise general oversight over research, education and outreach. The National Sea Grant Review Panel (NSGRP) had the responsibility of reviewing the NSGO and offering advice for conduct of the NSGCP. The NSGO solicited omnibus proposals from each individual Sea Grant program. The omnibus proposals included project proposals for individual research, outreach, and education projects and associated management proposals for implementation of program activities for the upcoming funding cycle. Funding levels for the omnibus proposals were based on the peer reviews and NSGO evaluations. Individual Sea Grant program directors then operated within the limits of their omnibus award and nonfederal funding. Site visits were conducted every two years by a NSGO review team to evaluate the program management process. Although the individual Sea Grant programs were not assured "base" funding (i.e., stable level of annual funding to support program activities), changes in response to reviews were relatively small; in practice, funding was fairly stable from year to year. Most individual Sea Grant programs conducted peer reviews to identify research project proposals to include in their omnibus proposals.

THE TRANSITION: 1994–1998

In 1993, the National Research Council (NRC) conducted a review of NSGO oversight and evaluation of the individual Sea Grant programs (NRC, 1994). The NRC review explored the roles of NSGO, the individual Sea Grant program directors, the NSGRP, and their respective responsibilities for program review and evaluation. The resulting report, *A Review of the NOAA National Sea Grant College Program,* was released in 1994 and recommended:

> The review process for research proposals should be decoupled from the NSGO evaluation of state programs prior to the 1995 reauthorization. Standard scientific and peer review procedures should be implemented for all state [individual] Sea Grant programs. The review process and all aspects of program implementation, including administration, should be streamlined prior to FY 1996. NSGO should evaluate the success of each state program on a four-year cycle, using, in part, retrospective information on recent achievements, based on measures for each of the three areas of research, education, and outreach.

NSGO began implementation of these recommendations in 1995. Program review was decoupled from the review of project proposals and institutional program directors implemented a standardized peer review and selection process for project proposals submitted to their programs. Congress reauthorized the NSGCP in 1998 and codified many of the recommendations of the 1994 NRC report, particularly with regard to program evaluation (for more details see http://www.sga.seagrant.org).

The 1994 NRC report recommended that a certain level of core funding be provided to each individual Sea Grant program to support an ongoing program of research, education, and outreach as long as the program performed at an "expected level of performance." The NRC report also recommended that changes in overall program funding be linked to past performance, with new funds awarded to individual Sea Grant programs on a competitive basis determined by the program review and evaluation process.

To establish a process for program evaluation, the National Director tasked the Committee on Procedures and Operations with developing recommendations for the protocol, criteria, and scheduling of a process for reviewing the individual Sea Grant programs (Copeland et al., 1997). The list of recommendations included: (1) a four-year cycle of external program reviews (25 percent of the programs each year) and ongoing assessment of the program by NSGO throughout the four-year period; (2) that the evaluations be framed in the context of a well-developed strategic plan and agreed to by the individual Sea Grant program director and the NSGO program officer, based on input by identifiable program advisors representing program constituents and institutional representatives (e.g., NSGO, 2004b); (3) institutional implementation plans be developed on a two-year cycle; (4) each individual Sea Grant program devise an internal review process to identify progress relative to strategic plan objectives; (5) regular progress reports, written by the individual Sea Grant programs, be provided to the assigned NSGO program officer; and (6) Topical Advisory Team (TAT) assessments be organized by an individual Sea Grant program director and NSGO program officer to address specific concerns that might arise during the review cycle. It is not clear how fully or uniformly these recommendations were implemented across the entire program. As emphasized in this report, strategic planning continues to be an area of concern with regard to program evaluation and the level of interaction between the NSGO and the individual programs.

PROGRAM REVIEW: 1998 AND BEYOND

Beginning in 1998, the NSGO implemented a quadrennial program review process recommended by Copeland et al. (1997). The first round of quadrennial reviews—Cycle 1—began in 1998 and was completed in 2001.[1] The second round of quadrennial reviews—Cycle 2—began in 2003 and will be completed by the end of 2006. While the basic framework of the quadrennial Program Assessment Team (PAT) reviews has been retained throughout Cycle 1 and Cycle 2, specific details of the program review process have been modified pursuant to internal and external reviews (e.g., Toll et al. 2001; Duce et al. 2002; Kudrna et al., 2005)[2] and congressional directives (33 U.S.C. 1121-1131; see Appendix H). The following "time line" gives key events (Table 2.1).

---

[1]The number of Sea Grant programs evaluated in Cycle 1 was 29, and the number in Cycle 2 was 30. In Cycle 1, Maine and New Hampshire operated and were evaluated as a single bi-state program. Before the start of Cycle 2, the joint Maine/New Hampshire program spilt into two programs that are now evaluated separately. There are currently 30 Sea Grant programs located in all of the coastal and Great Lakes states except Pennsylvania and the U.S. Commonwealth of Puerto Rico, with 3 additional programs in development stages.

[2]The Kudrna et al. (2005) review was released during the NRC study (November 2005), too late to assess the NSGO response for inclusion in this report. The summary of the Kudrna et al. review is reprinted in Appendix J.

Table 2.1 Timeline of Key Events Cited in This Report

Original Review Process	
1990	
1991	
1992	
1993	• NRC Study Begins
1994	• NRC Report 1
1995	• NSGO Response
1996	
1997	• NSGO Response • Copeland, Griswold, Fetterolf Report
Cycle 1 Reviews	
1998	• National Sea Grant College Program Reauthorization Act of 1998 (P.L. 105-106)
1999	
2000	
2001	• NSGRP Review of Cycle 1: *Review and Recommendations: Sea Grant Program Evaluation Process* ("Toll Report'")
Revisions to Review Process	
2002	• NSGRP Report: *Building Sea Grant: The Role of the National Sea Grant Office* ("Duce Report") • NSGO Response • National Sea Grant College Program Act Amendments (P.L. 107-299)
Cycle 2 Reviews	
2003	
2004	
2005	• NRC Study Begins • NSGRP Program Evaluation Committee Report: *Review and Recommendations: Sea Grant Program Evaluation Process* (Kudrna et al., 2005)
2006	• NRC Report 2

Following Cycle 1, the Toll Committee (see Box 2.1), named by the NSGRP, evaluated the procedures and made recommendations for modifications to address a variety of issues raised during Cycle 1. The NSGO subsequently made numerous changes in the details of the review and evaluation process (NSGO, 2005c) that were implemented in Cycle 2 (Duce et al., 2002; see Box 2.2). Differences in the criteria and evaluation processes under Cycle 1 and Cycle 2, and perhaps more importantly, lack of independent assessment prior to the implementation of these changes to establish a baseline, make it difficult, if not impossible, to directly compare the effectiveness of the evaluation processes used in each of the two cycles or to specifically tie improvement in the individual programs or the program overall, to changes in the evaluation process.

Box 2.1
The "Toll Report"

The National Sea Grant Review Panel's Program Evaluation Committee, chaired by John Toll, was charged in December 2000 with reviewing of Cycle 1 program reviews conducted pursuant to changes instituted in 1998. The resulting report, *Review and Recommendations: Sea Grant Program Evaluation Process* (often referred to as the "Toll Report", was published in October 2001. That report contained 40 recommendations, grouped into thirteen categories:

- NSGO Final Program Review and Merit Fund Allocation Process
- PAT
- Program Assessment Metrics
- Identification of Best Practices/Best Management Practices
- Public Notification of Upcoming Program Assessments
- Program Assessment Evaluation Criteria
- PAT Grades

Alternative #1: A Case for Eliminating Scores Assigned by the PATs
Alternative #2: Improved Standards for Program Assessment

- The Role of the NSGO Program Officer
- Effective and Aggressive Long-Range Planning
- The Biennial Implementation Plan
- Developing Guidelines for Self Evaluation
- TATs: Topical Advisory Teams
- Phase II of the Program Assessment Process

While the NSGO instituted many of the recommendations of the Toll Report (see NSGO, 2005c), several issues identified by that committee continue to be of concern, in particular the reliability of assessments conducted by different groups of individuals assessing different programs, the limited nature of constructive, ongoing interaction with NSGO staff, and the lack of a comprehensive planning process that can be implemented at both the local and national level.

Program Assessment Team (External Review)

The PAT[3] is a principal element of the evaluation process created by the NSGO in 1998. The PAT is a high-level "external" review team comprised of an NSGRP member as the chair, almost always an NSGRP member as vice chair, and 3 to 5 other members including an individual Sea Grant program director (of a program not under review), and other highly

[3]The PAT process will be discussed in some detail in chapters 3 and 4; it is introduced here simply to help the reader develop an understanding of the overall nature of the Sea Grant program.

Box 2.2
The "Duce Report"

The National Sea Grant Office Review Committee, chaired by Robert Duce, was appointed by the NSGRP in early 2001 to "conduct a comprehensive review of the NSGO and how it serves its many stakeholders, including its university partners, NOAA, the Department of Commerce, and other federal agencies."

Released in 2002 the resulting report, *Building Sea Grant: The Role of the National Sea Grant Office* (often referred to as the "Duce report"), made several recommendations that were intended to strengthen the NSGCP by improving the strategic planning process, encouraging cooperation among the individual programs to address regional challenges, and clarifying the roles and responsibilities of the NSGO. Four overarching points were made with direct relevance to the motivation behind and means utilized by the NSGO to carry out oversight and evaluation of the various individual programs. Specifically, the report recommended that NSGO:

- Lead in developing a comprehensive strategic plan for NSGCP and a national Sea Grant agenda.
- Provide leadership in communicating the NSGCP agenda, the achievements, and the opportunities of Sea Grant to Congress, the executive branch, and the public.
- Streamline and better manage the myriad administrative details essential to the operation of the NSGCP.
- Continue to seek adequate funding to effectively carry out the functions of the NSGO utilizing the findings of this report.

The "Duce Report" is widely seen as having a positive impact on the process. While many of the highest order recommendations regarding were adopted, concerns about the ability of the NSGO to more fully and meaningfully engage in the network development process remain.

regarded scientists, educators, and administrators from academia, government, and industry. The PAT receives training (NSGO, 1998) and is guided by detailed procedural and evaluation criteria (NSGO 1998, 1999a, 2000, 2001, 2003a, 2004a, 2005a) prepared by the NSGO and the NSGRP, and compiled to create the PAT Manual. Based on those guidelines and the presentations and documentation provided before and during a 3- to 5-day site visit, the PAT prepares a report outlining its findings and rating the program's performance in a number of areas.

PAT Guidelines

The NSGO has prepared a detailed manual with criteria and procedures to guide the PAT review and evaluation (for most recent PAT

Manual see NSGO, 2005a). The PAT Manual also provides guidance to individual Sea Grant program directors for preparing a briefing book to assist the PAT in their assessment process. The PAT uses materials provided in the briefing book and during the site visit to assess the:

1. overall productivity and accomplishments of the program relative to its strategic plan and level of support (NOTE: in both Cycle 1 and 2, both the adequacy of the strategic plan and progress made in implementing it were evaluated simultaneously);
2. overall scientific strength (e.g., the significance of scientific advances, the rigor of planning and internal review processes, the level to which available university talent and resources have been brought to bear on program goals and objectives, success in meeting program goals and objectives, publications and other output);
3. outreach and educational productivity and effectiveness;
4. management team effectiveness in planning and meeting stated goals and objectives, and in providing overall leadership for the program;
5. use of internal linkages among program elements and the ability to integrate these elements to address priorities (e.g., research, education, extension, and information dissemination);
6. position and role in its academic setting;
7. linkages with other Sea Grant programs, state and regional academic institutions, state and federal agencies, and the private sector;
8. linkages to industrial and user groups; and
9. potential for growth, considering all the above.

These nine assessment areas provide the framework for a more detailed set of review and evaluation criteria and benchmarks that is included in the PAT Manual and which frame the scoring by the PAT members to generate the PAT overall score (see Box 2.3 for example of a Cycle 1 scorecard).

Although much of the site visit is public, it is standard practice for the PAT to meet privately throughout the entire site visit (days, evenings, and whenever else is possible). The PAT discusses the review and comes to agreement on the evaluation scoring, findings, and recommendations, and writes its conclusions in an initial draft report (the final version of this report is called the "PAT Report," and it is discussed in the next section). Before concluding the site visit, the PAT meets with the individual Sea Grant program director and institutional representatives to discuss preliminary findings.

Box 2.3
Cycle 1 Program Score Sheet
(Reprinted from NSGO, 2001, p. 17)

Evaluation Criteria and Benchmarks for Performance—Summary

I. EFFECTIVE & AGGRESSIVE LONG-RANGE PLANNING:

The most effective programs will use the strategic planning framework from the NSGCP as a basis for developing their own strategic plan based on needs at the state and local level as identified in collaboration with a constituency advisory group. Effective planning may also involve regional programs.

(10%) Rating__________

II. ORGANIZING AND MANAGING FOR SUCCESS (4 Criteria).

MANAGING THE PROGRAM AND INSTITUTIONAL SETTING: Sea Grant programs are located within or work closely with university systems that are sites of major research and administrative activity. Each program must be managed to maximize the recruitment of outside resources to address Sea Grant problems and issues, as well as to build capability in the university system to address coastal problems and opportunities.

MERITORIOUS PROJECT SELECTION: The program carries out a good peer review and evaluation process for research, education and outreach projects, and selects those which receive consistently high marks for merit, application, and priority fit. The review must take into account how well a prospective project targets an issue.

PAT Report and Program Directors Formal Response

The NSGO has adopted a review process that directs the PAT to provide the individual Sea Grant program director and the NSGO with a comprehensive written report within 30 days of the site visit. This report should contain:

- documentation of the program's strengths and weaknesses;
- specific recommendations for program improvement; and
- an overall evaluation using the evaluation criteria and benchmarks for performance in the PAT Manual.

After receiving this final written PAT report, the individual Sea Grant program director has a reasonable time (until January of the following

RECRUITING AND FOCUSING THE BEST TALENT AVAILABLE: Every Sea Grant program has a variety of talent available for program development. The best efforts will involve the best talent. The program must have mechanisms in place to identify and attract the best talent available.

MERITORIOUS INSTITUTIONAL PROGRAM COMPONENTS: It is imperative that research projects, advisory programs, communications and education activities, and management use state-of-the-art methods and work to advance their disciplines.

(20%) Rating__________

III. CONNECTING SEA GRANT WITH USERS:

Effective information transfer occurs most often when the end users are involved in the planning and development stages, the program has an extension process in the field, and there is a mechanism for follow-up with users. The program management team should interact at the state, regional, and national policy levels. At the university level, the Sea Grant program must occupy an appropriate administrative and leadership position and be involved in decision making.

(20%) Rating__________

IV. PRODUCING SIGNIFICANT RESULTS:

The program must be managed to produce significant results. A basic mission of Sea Grant is to integrate research and outreach to address and significantly impact the identified needs of its constituency and of the nation.

(50%) Rating__________

OVERALL PROGRAM RATING__________

calendar year) to respond in writing. Most program directors do so, particularly in response to findings or conclusions with which the director disagrees or for which the director has additional information or perspectives. The PAT report and the director's response become part of the permanent record for the individual program and serve as the basis for the NSGO Final Evaluation Review (FE)(discussed below). It is envisioned that the PAT report will also establish a baseline for subsequent PAT assessments.

The NSGO and the NSGRP continue to work on guidance and training for PATs, to improve the quality of the PAT reports, to ensure that they are effective in informing the FE, and in guiding the individual Sea Grant program director in making improvements during the next review cycle (NSGO, 2005c).

PAT Review Criteria

During Cycle 1 (1998–2001), PAT reviews were framed around four criteria weighted as follows:

1. Producing significant results	50 percent
2. Organizing and managing for success	20 percent
• Managing the program and institutional setting	
• Meritorious project selection	
• Recruiting the best talent available	
• Meritorious institutional program components	
3. Connecting Sea Grant with users	20 percent
4. Effective and aggressive long-range planning	10 percent

Four categories (criteria) were used in Cycle 1 program ratings (NSGO, 2001, p. 16):

1. **Excellent**—If benchmarks of "Expected Performance" are substantially exceeded, the program will be rated as excellent.
2. **Very Good**—If the benchmarks of "Expected Performance" are generally exceeded, the program will be rated as very good.
3. **Good**—A program which generally meets the benchmarks of "Expected Performance" should be given a rating of good.
4. **Needs Improvement**—A program which does not reach the benchmarks should be given a rating of needs improvement.

Changes were made in the PAT review criteria and benchmarks based on the Toll et al. (2001), the Metrics Committee recommendations (see Appendix B of NSGO, 2005a), and in response to congressional requirements in the National Sea Grant College Program Act Amendments of 2002 (P.L. 107–299). These changes were implemented in Cycle 2, beginning in 2003 (see Table 2.2). (Details are included in the PAT Manuals (NSGO 1999a, 2000, 2001, 2003a, 2004a, 2005a.)

Cycle 2 PAT reviews have utilized a ratings sheet (score sheet) based on the four criteria of Cycle 1, but with a far more detailed sub-criteria for each criteria. Programs are rated as "Needs Improvement," "Meets Benchmark," "Exceeds Benchmark," or "Highest Performance" for each benchmark. The PAT manual has a detailed discussion of each "sub-criteria" (this report uses the term "sub-criteria" to refer to what the 2005 PAT Manual calls "sub-elements") and benchmarks and the percentage weight for each are shown in Table 2.2. Here are the brief descriptions of the four benchmarks from Cycle 2 as printed in the PAT Manual (NSGO, 2005a):

1. **Highest Performance**—Performance goes well beyond the benchmark for this sub-element and is outstanding in all areas.
2. **Exceeds Benchmark**—In general, performance goes beyond what would be required to simply meet the benchmark for this sub-element.
3. **Meets Benchmark**—In general, performance meets, but does not exceed, the benchmark for this sub-element.
4. **Needs Improvement**—In general, performance does not reach the benchmark for this sub-element. The PAT will identify specific program areas that need to be addressed.

The Metrics Committee

Following the Toll Committee report and its recommendation for improved metrics to fairly and uniformly evaluate programs across time, the NSGO appointed a Metrics Committee to examine potential qualitative and quantitative indicators of program performance and to make specific recommendations. The report of that committee *Indicators of Performance for Program Evaluation* was issued in March 2003 (Metrics Committee Report is included as Appendix B in NSGO, 2004a, 2005a). Subsequently, NSGO incorporated its recommendations and many of those in the Toll Committee report regarding metrics for review and evaluation.

FINAL EVALUATION PROCESS

The final evaluation process is carried out in five consecutive days during what is termed the NSGO Final Evaluation Review (FE), generally held in February. Participants include the NSGO leadership, NSGO technical staff members, plus nonvoting participation by one or more (usually two) members of the NSGRP.[4] The review looks back over all programs that were visited by PATs during the prior calendar year (a single PAT cohort). The result of the FE is a summary letter from NSGO and a score upon which merit and bonus funding decisions are based.

The FE differs from the PAT review in several ways because more information, collected over time (1 to 4 years), is incorporated from NSGO assessment during the review cycle. According to Sea Grant program documentation and reports (discussed earlier in this chapter), the FE considers 7 or 8 programs simultaneously, thus providing a comparative perspective across programs, based on the following information:

[4]During the 2005 NSGO Final Evaluation Review, one member of the NRC review committee and one OSB staff member were included as observers.

Table 2.2 The Review Criteria Ratings Sheet ("Score Sheet") for Assessments and Reviews During Cycle 2

		Needs Improvement	Meets Benchmark	Exceeds Benchmark	Highest Performance
1. ORGANIZING AND MANAGING THE PROGRAM	**20 %**				
*Sub Elements**					
Leadership of the Program	6 %	______	______	______	______
Institutional Setting	4 %	______	______	______	______
Project Selection	2 %	______	______	______	______
Recruiting Talent	3 %	______	______	______	______
Integrated Program Components	5 %	______	______	______	______
2. CONNECTING SEA GRANT WITH USERS					
Sub Elements	**20 %**				
Engagement with Appropriate User Communities	15 %	______	______	______	______
Partnerships	5 %	______	______	______	______
3. EFFECTIVE AND AGGRESSIVE LONG RANGE PLANNING	**10 %**				
Sub Elements					
Strategic Planning Process	4 %	______	______	______	______
Strategic Plan Quality	4 %	______	______	______	______
Implementation Plan	2 %	______	______	______	______
4. PRODUCING SIGNIFICANT RESULTS	**50 %**				
Sub Elements					
Contributions to Science and Technology	10 %	______	______	______	______
Contributions to Education and Outreach	10 %	______	______	______	______
Impacts on Society, the Economy, and the Environment	25 %	______	______	______	______
Success in Achieving Planned Program Outcomes	5 %	______	______	______	______

* The text of this report refers to "sub-elements" with the term "sub-criteria".

SOURCE: NSGO, 2005a , pp. 16 and 19.

1. The documentary material used in the FE includes:
   - The PAT report along with the institution's response,
   - The program's strategic plan/implementation plan,
   - Annual progress reports,
   - Information on major accomplishments,
   - Trip- and peer-review-panel reports by the Program Officer (if any),
   - Topical Advisory Team reports (if any),
   - Sea Grant funding information,
   - Other material deemed to be relevant by the Program Officer,
   - Four-year project-by-project report on Sea Grant funding, and
   - Copies of the PAT briefing books and omnibus proposals.
2. Insights (provided by the NSGO staff) into each program's performance, management, and results based on interactions with the programs over the entire four-year review period; and
3. Insights (provided by the NSGO staff) into the contributions of the individual programs in support of the total National Program. For example, whereas the PAT would evaluate program management in terms of the results and output, the NSGO would add considerations of how well the program supported NSGO and national initiatives, and the degree to which the program functioned and identified itself as part of the national Sea Grant network. Collaborative efforts among the programs are given credit.

The manner in which these FE deliberations are carried out, with subsequent distribution of merit funds, was first described in a policy memorandum dated April 22, 1999, sent to the Sea Grant directors from the National Director (see Appendix D). This initial process was used until 2003 when, as a result of 2002 congressional action, the merit and bonus funding procedures were modified. Draft revisions of the policy document were circulated in 2004 for comment and the new version was promulgated April 8, 2005 (NSGO, 2005c).

The bulk of the FE review week is spent, about half a day at a time, considering each of the individual programs reviewed the previous year. The program officer for each program begins with a formal presentation, following a common template, describing various aspects of the program being considered. This is followed by a detailed discussion, facilitated by the National Director, of the performance of the program in each of the evaluation criteria listed in the PAT Manual (note that the four criteria used by the PATs in Cycle 1 were subdivided into 14 criteria for Cycle 2, see sub-criteria in Table 2.2). These criteria-based discussions are the foundation for scoring programs for purposes of merit and bonus funding.

At the conclusion of the discussion of each criterion, the group votes to assign one of the four ratings to that criterion (as stated earlier in this chapter, ratings were slightly different from Cycle 1 to Cycle 2. ["good" vs. "exceeds benchmark," etc.]). The group vote in Cycle 1 was phrased in terms of agreeing with the original PAT rating or assigning a higher or lower evaluation (Schuler, 2005). In Cycle 1 and part of Cycle 2, a simple majority vote (more than half of the votes cast) was required to change a PAT rating. In response to concerns about the impact of assigning a score during the FE that differed from that provided by the PAT, steps were taken to revise the NSGO's role in the rating of individual programs. In 2005, this voting process changed to require a two-thirds majority (more than two-thirds of the votes cast) to assign a score different from the PAT rating. In effect, this reduces the role of the NSGO in rating individual programs; the significance of this change will be revisited in chapters 3 and 4.

The final day of the of the FE week is spent reviewing the cohort of programs assessed by a PAT the previous year criterion by criterion, primarily emphasizing instances in which the FE rating differed from the PAT rating, but including instances in which comparative judgments among programs might lead to different ratings for programs discussed in the early part of the week. A significant function that takes place in the FE is the assignment of numerical values (i.e., scoring) to the ratings for the 4 criteria or 14 sub-criteria (for Cycle 1 or Cycle 2 respectively).

The ratings assigned to each criterion are then converted to a numerical equivalent on a four-point scale, so that the highest rating is given a 1.0 and the lowest a 4.0. Thus, the poorer the performance, the higher the score. These scores for the various criteria are then combined using the weightings described in the PAT manual to create an overall program score (NSGO, 2005a, p. 16).

Program Performance Rating

Based on the FE score, all 30 programs are divided into 4 categories. The scores of programs in each of the four categories vary considerably in age. Although one-quarter of the scores are at least three years old and three quarters of the scores more than a year old, the 1999 NSGO memorandum makes no mention of numerical scores but defines the "rating categories" stating: "Ratings are based on grading of the same four categories as the PAT evaluations" (NSGO, 1999b, p. 4).

The 2005 version states "The NSGO final rating for the program is determined by locating a program's score along a fixed four-category rating scale for merit funding and a variable two-category rating scale for bonus funding" (NSGO, 2005b, p. 7). Historically, the actual scores were kept confidential from individual Sea Grant programs, but starting in

2004, programs were informed of their final scores, and any differences between the PAT and FE were explained. Beginning in 2005, each program was given a summary of the overall performance boundaries and number of programs within each boundary, but individual program scores remain confidential. As a matter of practice, category 4 is rarely assigned; such an assignment triggers special interactions to improve program management. The relevant policy memoranda are included in appendixes D and E in this report.

Final Report, Ranking, and Allocation of Funds

Following the FE, the National Director prepares a final report for each program and transmits it as a letter to the individual Sea Grant program director. This final report summarizes the evaluation results for each program for each of the four major evaluation categories. Even though the Cycle 2 evaluations subdivide the 4 categories into 14 sub-criteria, the final report for Cycle 2 reviews does not address the 14 sub-criteria separately in an effort to maintain focus on the four major categories.

In Cycle 1, and for the first two years of Cycle 2, it was the practice of the National Director not to include a final rating (e.g., Highest Performance, Exceeds Benchmark, etc.) in the letter to the individual Sea Grant program director. Instead, the Program Officer would inform the individual program director that the program had been assigned to a merit category.

As discussed earlier, starting in 2005 (part-way through Cycle 2), the letter from the National Director to the individual programs included the ratings for each of the 14 sub-criteria and specified into which merit category the program had been placed.

During the Cycle 1 and partial Cycle 2 reviews, the NSGO director's letter informed the individual Sea Grant program director of the actual amount of merit funding awarded (NSGO, 2003b). For Cycle 1 this funding was in two parts: one remained fixed until the next review was completed (4 years) and a second, smaller amount varied from year to year as additional programs were evaluated and additional programs entered or exited the top three categories. In Cycle 1 this merit funding was based on the score and the number of programs with similar scores, not on the relative position of the program in the overall ranking.

Beginning in Cycle 2, evaluation, ranking, and the merit award process became more complicated. Not only did the number of evaluation criteria increase from 4 to 14 (as sub-criteria were specified), but Congress (P.L. 107-299) mandated a competitive ranking formula based on five categories, with no more than 25 percent of the programs ranked in the top category (Merit Category 1), and no more than 25 percent could be ranked in the second category (Merit Category 2). NSGO responded to this formula by creating a ranking formula that contained, in essence, a total of six categories: Merit Category 1 (the top-ranking category) was subdivided into 1A (containing the top seven programs, just under 25 percent), 1B (containing the next seven), and 1C (containing the remainder of programs in Merit Category 1). Merit categories 2 and 3 remained as required by Congress, and Category 4 represented the sixth category. To date, no program has been found to perform so poorly as to be assigned into Category 4. While the overall category (e.g., Category 1, Category 2) to which a program is assigned (for determination of merit funding) remains unchanged during the period between reviews, the scored programs ranked in the category subdivisions 1A, 1B, and 1C change yearly as additional programs are reviewed and relative rankings within the category change.

For funding allocation, the amount of the merit award to a program remains unchanged throughout the period before the next PAT review. However, an additional bonus fund distributed to programs in sub-categories 1A and 1B may change annually.

Details of the method of ranking and allocating the merit and bonus funds are given in a Policy Memorandum on NSGO Final Evaluation and Merit Funding (NSGO, 1999b), revised in 2005 (see Appendixes D and E). An example of how the Cycle 2 funding allocations might play out is shown in Figure 2.1. The fact that the allocations are for specific dollar amounts unrelated to the size of the individual program's core funding makes the reward, and changes to it, much more significant for the smaller programs (see Chapter 3 for more discussion).

CONCERNS WITH THE PROGRAM EVALUATION PROCESS

Given the complexity and diversity of individual Sea Grant programs and the complex funding strategies of each program, the National Director, NSGO staff, and the NSGRP have developed a detailed process resulting in meaningful review and evaluation. There are, however, several shortcomings that could be rectified to make the overall process more effective.

Since the reauthorization of the program in 2002, program evaluation within Sea Grant has evolved to serve two, theoretically related purposes. The 2002 amendments redefined the purpose of evaluation from simply gauging and encouraging improvement in individual programs to rating programs, "relative to each other for the purpose of determining performance-based funding." These dual purposes are related insofar as competition for funds serves as an incentive to the individual programs to

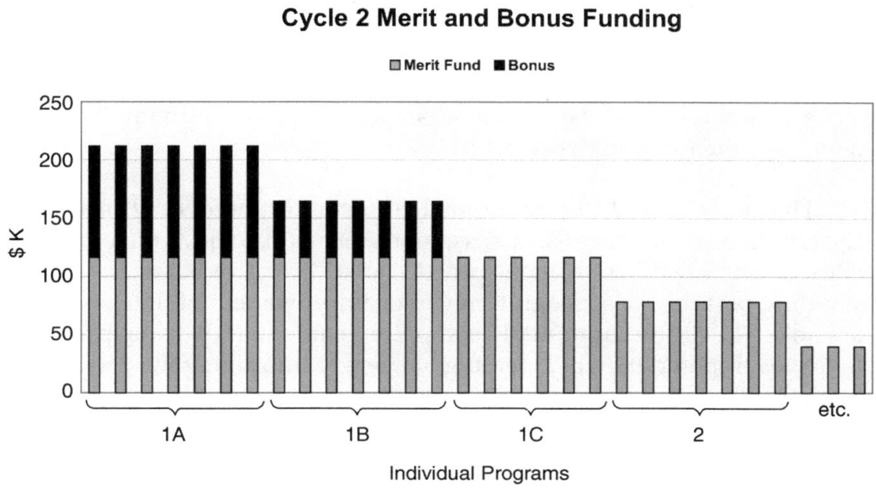

FIGURE 2.1 How a hypothetical $3 million merit funding pool with a $1 million bonus funding pool might be allocated among 20 programs that have been ranked in Category 1. It should be noted that it is possible for a particular category to have no programs assigned to it. For example, if there were 14 or fewer Category 1 programs, the third group (1C) would have no programs assigned to it (NSGO, 2005b, p. 13).

improve. However, an evaluation process that is well designed for identifying areas and mechanisms for program improvement may be inadequate for ranking programs. A process whose foremost purpose is to rank programs may do a poor job of encouraging aspects of program improvement.

The process must be balanced so that efforts to achieve one objective do not undermine efforts to achieve the other. Furthermore, Sea Grant is often considered a network or partnership, thus the process must balance efforts to improve the effectiveness of individual components against improving the effectiveness of the network or partnership as a whole. An "ideal" assessment process would include the following characteristics:

- Credible (uses professionally recognized methods or "best practices" within the field)
- Reliable (results should be reproducible)
- Meaningful (criteria, benchmarks, and indicators should reflect characteristics of an effective program defined in terms of national, regional, and local benefits)

- Cost-effective (cost of effort, in terms of human and fiscal resources should not exceed a reasonable fraction of the annual budget of individual programs or the network as a whole)
- Comprehensive (should assess effectiveness of individual components as well as the network itself)

The process could be fine-tuned to focus on the overall objectives. Designing or modifying the assessment process to achieve the characteristics of an "ideal" assessment process would require balancing of outstanding performance recognition versus improvement of the network as a whole. The desire to stimulate competition among individual programs must be tempered to avoid creating barriers to improving the program as a whole. This could be achieved if emphasis were placed on rewarding the outstanding performer rather than on stigmatizing the acceptable performer. Approaches for achieving such balance, based on detailed analysis of the current process are explored in chapters 3 and 4.

3

Critique of the Periodic Assessment Process

Chapter 2 described the evolution of procedures that are currently used to assess the performance of individual Sea Grant programs. As mentioned, the character of the program assessment process is dominated by periodic aspects: the quadrennial visit of a Program Assessment Team (PAT), followed by the National Sea Grant Office (NSGO) Final Evaluation (FE) Review.

This chapter presents a critique of this periodic portion of the assessment process. The first section discusses the documents that provide guidance on which the review procedures are based and carried out. The second section provides a critique of the primary element—the onsite review carried out by the PATs. The third section examines the FE process carried out by NSGO staff during an intensive review of the programs that make up the most recently reviewed PAT cohort (7 to 8 programs in a given year) and results in a final evaluation letter from the National Director to the individual Sea Grant program director. The final section considers the assessment process as a whole, its use in assignment of merit and bonus funding, and proposes a realignment of functions intended to strengthen the program overall.

GUIDANCE DOCUMENTS

The periodic assessment process follows instructions provided in two guidance documents: the PAT Manual (NSGO, 2005a) and the Policy Memorandum (NSGO, 2005b). The PAT Manual provides detailed instructions on conducting the PAT visit. The Policy Memorandum outlines the structure and function of the FE process, including details on how funds are allocated based on program scores derived from the PAT visits and reviewed during the FE. These two key documents do not appear to be as well known among the relevant parties as they should be.

Observation of the various PAT site visits taking place in 2005 made it clear that while the individual Sea Grant program directors were familiar with the PAT Manual and policy memoranda, some other staff involved with preparing background documents and briefing reports for the PAT had not seen the PAT Manual even by the end of the PAT visits. Also, there appears to be some significant confusion about the FE process, despite the fact that relevant policy memos (available in the administrative information portion of the NSGO web site at http://www.seagrant.noaa.gov/other/admininfo.html) answer the vast majority of the most frequently posed questions. Thus the most frequently raised concerns do not appear to reflect a lack of specificity or availability of these documents, but rather a lack of familiarity with them. The NSGO needs to disseminate the contents of the documents more actively and broadly through a process that involves active and personal explanation of the periodic program assessment process with staff as well as directors of the individual Sea Grant programs. The individual program directors should disseminate, to their staffs and all others who will be taking part in the review, the contents of these documents, particularly the PAT Manual. The result would be a more satisfying PAT site visit for all concerned.

The more detailed of the two documents, the PAT Manual, identifies the review criteria, the benchmarks used to describe the expected level of performance in a particular area (such as program organization and management), and the indicators used to help assess the outcomes or impacts of the individual program against the benchmarks (see Appendix G). These set the standard for performance and provide a basis for rating the individual Sea Grant programs in relation to established expectations. The specific wording of these items has evolved over time, under intense scrutiny and regular feedback from PAT members, individual Sea Grant program directors, the National Sea Grant Review Panel (NSGRP), and program officers.

Throughout the history of the process, there have been four main criteria for assessment reflecting the breadth of the activities for which each program is responsible: (1) organizing and managing the program[1] (20 percent); (2) connecting Sea Grant with users (20 percent); (3) effective

[1]This criterion, originally named "Organizing and Managing for Success" in Cycle 1, was renamed in Cycle 2.

and aggressive long-range planning (10 percent); and (4) producing significant results (50 percent). The weight given to each of these criteria has remained constant throughout Cycle 1 and Cycle 2, although in Cycle 2 each of the four criteria categories is subdivided into 2 or more sub-criteria with individual weightings—14 sub-criteria in all. The 4 major criteria are balanced well between evaluation of potential to perform and performance itself, but focus extensively on how the program performs at a local level (this aspect will be revisited in Chapter 4).

For each of these criteria one or more benchmarks are provided. The benchmark is a description of what constitutes acceptable performance. For example, the sub-criterion "Institutional Setting and Support" accounts for 4 percent of the overall score and appears under the criterion category "Organizing and Managing the Program." The "expected performance benchmark" is:

> The program is located at a high enough level within the university to enable it to operate effectively within the institution and externally with all sponsors, partners, and constituents. The institution provides the support necessary for the Sea Grant program to operate efficiently as a statewide program (NSGO, 2004a).

The internal complexity of each benchmark leaves room for the evaluators (PAT members) to weigh the different elements appropriately for the program in question. The evaluators are also asked to take into account indicators of performance and a list of "suggested considerations." Asking knowledgeable evaluators to incorporate such diverse sets of information into an overall score is a standard part of assessment processes in research organizations. In particular, using quantitative indicators to inform qualitative judgments, as the Sea Grant evaluation process does, is widely considered the best use of performance criteria. The current Sea Grant benchmarks have variable formats and sometimes mix management and results concepts in the same benchmark (e.g., under "effective and integrated program components" the list of expected performance benchmarks includes "research results are consistently reported in peer-reviewed publications"), but are by and large quite well done and are consistent with the goal of assessing, and thus guiding, performance of individual Sea Grant programs.

The use of performance criteria in underpinning subjective evaluations is treated in Appendix B of the PAT Manual. Much of that treatment is of a general nature, defining and recommending the use of performance criteria to inform the review process and contribute reliably to comparability among different PATs. This is followed by a list of possible indicators related to the four broad criteria, on which the overall review process is based. See Box 3.1 for list of indicators in the 2005 PAT Manual.

Box 3.1
Indicators of Performance Organized in 4 Categories
Reprinted from 2005 PAT Manual (NSGO, 2005a)

1. INDICATORS FOR PROGRAM MANAGEMENT

Managing the Program—*Response to previous PAT recommendations; Management Team composition and responsibilities; Percentage time Director and staff devote to SG (FTEs [Full Time Equivalent]); Advisory Boards membership and function (expertise, meeting schedule, recommendations,* meeting agendas, attendance, diversity, and turnover); Staff structure, interactions, and physical location in state

Institutional Setting—*Setting of the program within the university or consortium organization and reporting structure*; Program infrastructure (space, equipment, available resources)

Project Selection—*Process to develop RFP [Request for Proposal] priorities; Preproposals and proposals submitted, and institutions represented / institutions available in state; Review process including composition of panels;* RFP distribution; External peer review (numbers and quality), ratings/ scoring analysis, quality of feedback to PI's; Conflict of interest policy and practice; Time from submission to decision; Technology support for submission and review process; Feedback from PIs and/or institutions

Recruiting and Focusing the Best Talent Available—*New vs. continuing projects and PI's [Principal Investigators]; Recruitment of PI's/institutions; Relative success of home institution; Success in national competitions; Regional/multi-program projects*; Multiinvestigator projects; Leveraged funding in projects

Institutional Program Components—*Integration of outreach and research program elements; Core Federal and matching funds (last 8 years) and distribution among program elements; Leveraged funding from partners (NOAA, other Federal, State and local) for the program; National competition funding (NSIs [National Strategic Initiatives], pass through awards); Additional Program Funding through grants, contracts and development activities;* Leveraged funding from partners (NOAA, other Federal, State and local) for PIs

2. INDICATORS FOR CONNECTING WITH USERS

Constituent Involvement in Planning—*Local business and stakeholder needs surveys*; User feedback (mechanisms and tracking)

Contact with Appropriate User Communities—*Leadership by staff on boards and committees*; Informational meetings/training sessions held and number of participants; Individual consultations with clients/users; Involvement with industry (number of businesses aided); Demographics of contacts and efforts; Requests for information

Partnerships—*Effective local, regional and national interactions/collaborations including with NOAA programs*

Implementation—*Number, list and diversity of products produced print, audio, video, web, etc); Internal evaluation processes for products and programs; Staff and product awards*; Targeted audience and evaluation for all products; Media interest (calls, "experts quoted," press clippings); Use of products for public education (classroom enhancement, curriculum development); Relationship of products to other SG program elements; Numbers of teachers and/or students using Sea Grant materials in curriculum

3. INDICATORS FOR PLANNING

Planning Process (Input)—*Stakeholder and staff involvement (numbers and duration) and integration of input into planning; Transparent priority-setting process;* Endorsement by Advisory Board; acknowledgement by University Ongoing monitoring of plan and reassessment of priorities

Plan Quality (Goals, Objectives, etc.)—*Short to long-term functional and management goals established*; Demonstrated link from state to national priorities

Plan Implementation (Strategy and Tactics)—*Distribution of investment effort to meet strategic plan priorities; Identification of short to long-term benchmarks; Work plan developed for integration of program elements*; Program development and rapid response procedures and strategies to meet emerging issues; Evaluation process

4. INDICATORS FOR ACHIEVING SIGNIFICANT RESULTS

Contributions to Science and Engineering—*Number and list of publications (journal articles, book chapters, reports, etc); Invention disclosures and patents*; Technologies and tools developed; Theories or approaches accepted widely; Number and list of presentations by PI's; Citation analysis for selected projects

Contributions to Education—*Numbers of graduate and undergraduate students supported, including fellowships and internships; Sponsorship of education programs and target audience participation*; Changes in behavior of target audiences; Numbers of theses completed; Tracking of graduate students after Sea Grant support

Socioeconomic Impact—*Descriptions of the most important impacts; Positive environmental impacts and economic benefits resulting from changes in behavior of individuals, businesses, and institutions*; Businesses and jobs developed after contact; Best management practices developed in response to extension involvement

Success in Achieving Planned Program Outcomes—Self-assessment

Under each of the four criteria the required and suggested indicators are organized into three to five sub-groups. The large number of indicators (approximately 100) attests to the complexity of the review, but the organization of indicators into sub-groups provides a useful framework for understanding the most valued characteristics of an individual Sea Grant program.

An essential contribution made by the study of performance criteria is to improve the efficiency of activities such as the Sea Grant review process. Under certain circumstances careful analysis may show that an approach with 5 criteria would lead to as reliable a result as one using 10 criteria, or that the use of 50 indicators may be as useful as 100. In the present case the argument is made below for the reduction of the number of criteria from the current 14 to a significantly smaller number but with the addition of a criterion that would assess activities to strengthen the ability of programs to cooperate on regional or national scale issues.

Determining the most appropriate number of indicators is not simple. Reducing the number of indicators might be advantageous to reviewers when carrying out their tasks, but shortening the list of indicators might be a disservice to the individual Sea Grant program directors who must prepare briefing materials. The director's task is to anticipate and provide answers to questions that the reviewers might logically raise. The indicators listed in the 2005 PAT Manual (see Box 3.1) all appear to represent relevant questions that could reasonably be expected to come up during the review. Because no one wants to be caught off guard during a review, these indicators aid in preparation.

While the instructions for both Cycle 1 and Cycle 2 asked reviewers to give four levels of rating, the labels and instructions varied somewhat between the cycles (see example of score sheets in Chapter 2—Box 2.3 and Table 2.2). In the current instructions, the reviewers are asked to assign one of four ratings: needs improvement, meets benchmark, exceeds benchmark, or highest performance. Some description is provided for each rating level, although there can be considerable subjectivity involved in distinguishing between the "exceeds benchmark" level (described as "in general goes beyond") and the "highest performance" level (described as "goes well beyond and is outstanding in all areas"). In addition, the definitions of some benchmarks include superlative language (e.g., exceptional talent) that would make it difficult to distinguish benchmark performance from the "highest level". Further fine-tuning in the rating instructions is possible and advisable, but no grading system will ever eliminate subjectivity entirely.

The earliest set of instructions to PATs had one benchmark for each of three criteria and four benchmarks in a fourth criterion. Evaluators provided just four ratings, one for each performance criterion. In the PAT manuals for Cycle 2, there are 14 sub-criteria,[2] because each of the original four major criteria used in Cycle 1 were subdivided into at least two sub-criteria. Evaluators provide a rating for each of the 14 sub-criteria. Each criterion still carries its own weighting, which now ranges from 2 percent to 25 percent, and the final score is the sum of the products of the 14 ratings and weights for each criterion.

This subdivision into 14 weighted sub-criteria was not recommended by any of the major committees that have examined the process. Nor is there evidence to suggest that 14 weighted sub-criteria provide a more accurate assessment of program performance than a smaller number of criteria. The 14 weighted sub-criteria may also increase the perception that individual Sea Grant programs now have to "teach to the test," that is, that the very specific criteria skew behavior. Performance measurement systems always have the effect of orienting behavior, and good ones are carefully balanced to make sure that all the kinds of behavior that are actually important are included. Consideration should be given to reducing the number of weighted criteria to be assessed in the future, but implementation should be postponed until the beginning of the next cycle of program review (the current review cycle will conclude in late 2006). With only 4 to 6 broader criteria, weighted to reflect a balance between the production of meaningful results, outreach and education, and planning, organization, management and coordination among programs, the PATs would be able to form more holistic judgments of overall program performance.

All parties involved in the review process have been concerned with how PATs made up of different groups of volunteers could rate different programs in consistent ways (e.g., would the same actions in two programs receive different grades if evaluated by different visiting PATs). In an effort to characterize the problem, a simple statistic was calculated to measure overlap among PATs over the course of a four-year cycle. For each cycle, the proportion of pairs of PATs that shared at least one member was calculated. Although this statistic could be calculated for a given time period, the statistic for overlap within a given cycle is the most relevant, given that a program is ultimately ranked against all 29 of its partners. The results for both Cycle 1 and the partially completed Cycle 2 show a low proportion of overlap, 0.24 and 0.30 respectively. In addition, overlaps with more than one person were rare. The average numbers of shared members were 0.26 and 0.35, for Cycle 1 and 2 respectively. Thus,

[2]The term *criterion* has been used differently at various points in the evolution of the evaluation standards, but currently refers to these fourteen areas.

although recent efforts to improve the reliability of PAT reviews by increasing the overlap among PAT membership appears to have had some effect, the actual effect is still relatively low.

The Sea Grant program assessment process has taken several steps to attempt to achieve reliability in ratings. First, the NSGRP—a standing committee from which PAT chairs are selected—is represented in the FE process and provides continuity and broad assistance in PAT guidance and training, including work on providing grades in consistent ways across PATs. Second, the NSGO tries to have some overlap in the membership of PATs, so that someone is present at the PAT who can do comparisons across at least two programs between Cycle 1 and Cycle 2. This does not, however, address reliability among reviews within a cycle. Third, the benchmarks are designed to provide a standardized comparison point in each of the four rated criteria. Both PATs and NSGO staff use the same criteria, sub-criteria, benchmarks, indicators, and ratings instructions in their evaluations. Finally, the last day of the FE is devoted to comparing grades across programs and adjusting them to reflect differences in performance consistently. This final step, though necessary, underscores the importance of NSGO being well positioned to independently and credibly evaluate the individual programs across the breadth of the entire program.

The April 8, 2005, Revised Policy Memorandum on NSGO Final Evaluation and Merit Funding (NSGO, 2005b) from the National Director to the individual directors moves significantly toward the goal of improving the transparency of these processes. It carefully describes the information that is considered in the FE, the procedures by which the process is carried out, and the ways in which this review differs from and parallels the PAT process. It also describes in detail the manner in which the merit and bonus decisions are made but does not specify how the performance criteria categories and relative standings are defined in terms of the resulting numerical scores.[3]

PROGRAM ASSESSMENT TEAM VISIT

Currently, the site visit by the PAT is the defining event of the periodic review process. The concepts of program review and accreditation are well established in the academic community and among granting agencies. The one aspect that distinguishes these events from most simi-

[3]The description of how qualitative ratings in the FE are converted into numerical values can be found in NSGO, 2005b, also included here as Appendix E.

lar activities is the element of competition. Most reviews of ongoing programs are carried out to determine whether the program is doing well against some set of mutually agreed upon goals. While this is true for the PAT visit and report, an additional element of competition was formally introduced in response to the National Sea Grant College Act Amendments of 2002 (P.L. 107–299).

The competitive process is directly affected by differences among the personnel in the various PATs. While NSGRP activities, the guidance documents and previsit training are all conscientiously applied, further improvement would result from measures that would facilitate the overlap of personnel among several review teams. Overlap is essential both within cycles and between cycles, as discussed earlier in this chapter.

Many of the visits have required four or five days of project review, field trips, and program presentations, raising concerns about the financial cost and the demand such efforts place on PAT members, reducing their ability or desire to serve on more than one PAT. Shortening the PAT visit could save expenses and time devoted to preparation and conduct, and the expense of clientele and principal investigator appearances before the PAT. Because Sea Grant is a partnership between NSGO and the institution, the PAT visits are often designed to satisfy the host institution's requirement for periodic external review of academic programs. Consequently, the desire to shorten the length of the PAT visit should be tempered by the need to be responsive to the individual program needs. The PATs need to understand the individual Sea Grant program's manifold dimensions.

In March 2005, the NSGO added a new section to the PAT Manual (NSGO, 2005a) entitled "PAT Preparation, Structure and Cost Control." Under this section, the NSGO provides suggestions for ways to minimize the costs of the PAT visit, without reducing the PAT's effectiveness. The content of that section is summarized below:

- Field trips should be used sparingly, and when appropriate, sessions with formal presentations can substitute for field trips.
- Expensive venues should be avoided.
- Expensive social events and dinners are not expected.
- Receptions can be combined with poster sessions.
- Quality of briefing book depends on content, not on glossy publications.[4]

---

[4]The PAT Manual (NSGO, 2005a) includes an appendix on "Guidelines for Program Assessment Briefing Books" that recommends brevity in briefing book preparation.

- Use of CD-ROMS for auxiliary materials is encouraged.
- Use conference calls, web and video conferencing where appropriate to reduce travel expenses and engage important community leaders who may not be able to attend in person.

The NSGO should be commended for encouraging reducing the costs and fanfare in its 2005 PAT Manual. Putting a program in its best light can be achieved more effectively by providing an easily digested amount of well focused, content rich material. Thus, another way of reducing preparation time of the site visit is for the Program Officer, the individual Sea Grant director, and the PAT chair to have some flexibility in deciding how to organize the visit. Perhaps, it would help to highlight certain issues or activities, while still using the performance criteria consistently. The success of shorter and more focused PAT site reviews will depend, in part, on increased engagement and continuous oversight by the Program Officer and the ability to identify and focus on important program areas, as discussed and recommended in Chapter 4 of this report.

More efficient and shorter PAT site visits could allow NSGO to conduct site visits to half the programs in one year and the other half the following year. This might make it easier for PAT members to participate in several site visits and provide better comparison among programs. At the end of two years, all programs can be more effectively compared and ratings of program performance would be more comparable.

FINAL EVALUATION PROCESS

The FE and merit and bonus funding process is introduced in Chapter 2 of this report, and it is described in the National Director's memoranda of April 22, 1999, and April 8, 2005 (see appendixes D and E). The FE process has been the subject of frustration for some individual Sea Grant program directors who characterize the FE as "lacking transparency" or as a "smoke-filled room" event, where program scores are changed for reasons that are unknown or not understood by the individual Sea Grant program directors. A significant cause for this perception appears to be poor communication in several areas. In one exchange of letters between an individual Sea Grant program director and the National Director, it was clear that the Sea Grant program director was not aware of the 1999 Policy Memorandum describing the FE process. Primarily prompted by the introduction of the rating and ranking process mandated in the 2002 reauthorization (P.L. 107–299) and by the implementation of this new process, the 2005 Policy Memorandum was written in an attempt to clarify the FE process. The NSGO sent out successive drafts in 2004 for comment and made significant revisions based on comments received. However, because the final 2005 Policy Memorandum was not available until after the FE week (a 5-day meeting usually held in February), the degree to which it will clarify the process and reduce tensions is not yet known.

The letters that the NSGO director sends to the individual Sea Grant program directors at the conclusion of the FE process may also contribute to the perception of a lack of transparency. Although in many respects these letters are quite similar to the letter sent to the individual Sea Grant directors after the PAT report, they differ in one important way. In early portions of Cycle 2, the comments in the final letter are compressed from the 14 criteria used by the PAT into the four larger categories, and do not include any final score. This issue was addressed by a procedural change in 2004, which led to the practice of including the final score in the FE letter.

Differing perspectives and program obligations of the NSGO and the individual Sea Grant directors, as well as insufficient communication and program liaison, appear to contribute to a tension that fuels the perception of lack of transparency and misunderstanding of the role of the FE. These tensions are understandable given a national program that is implemented by state and local directors and staff who are passionate about their work. Several actions discussed and recommended in this report, such as better NSGO communication with individual programs, increased program officer engagement, and more integrated strategic planning, could help to improve operational trust and respect among all program levels, thereby facilitating efforts to further improve the program and enhance its station within the community.

Credibility of PAT and FE Scoring Process

In Cycle 2, the number of criteria was increased to 14 over the 4 of Cycle 1. However, the 14 sub-criteria were simply subdivisions of the 4 major criteria used in Cycle 1; thus, the distributions of the FE and the PAT differences among the 4 broad categories can still be assessed. In the 8 reviews of the first year of Cycle 2 there were 2 disagreements in "Significance of Results" and 1 for "Connecting with Users" (combined carrying 70 percent of the ranking weight) as opposed to 9 disagreements in "Organizing and Managing the Program" and 11 in "Effective and Aggressive Long Range Planning" (recall there were multiple sub-criteria under each of the 4 criteria). This distribution was similar to differences seen in Cycle 1 and implies that in spite of the involvement of members of the NSGRP in both the PAT and FE processes as a communication link, there is often not a common view of program performance under these criteria.

In year two of Cycle 2 the procedure changed from requiring a simple majority to requiring a two-thirds majority for an FE rating to differ from the PAT rating and the number of disagreements dropped substantially. There were nearly 3 changes per program (2.875, a weighted average out of a total of 14 criteria and reported on a four-point scale, as they have been since 1998) in year 1 and only just over 1 change per program (1.14, also a weighted average out of a total of 14 criteria, reported on a 4-point scale) in year 2. The distribution changed as well with half of the disagreements being in the "Significance of Results" category. While this outcome is correlated, it is not necessarily causal. To fully understand the significance of this correlation, one would need to know how many changes were proposed but failed to win a two-thirds majority or how many changes were not proposed because they were unlikely to win a two-thirds majority.

While there were many differences between the PAT score and the FE score in both cycles, these differences were not predominantly either positive or negative[5] The mean overall score difference in Cycle 1 was 0.0047 and the mean overall score change in Cycle 2 was 0.0093. Because the mean overall score difference includes positive and negative differences, it does not provide a good representation of the typical difference between the PAT and FE score for individual programs. The mean absolute overall score difference is indicative of the typical magnitude of differences in PAT versus FE scores; in Cycle 1, the mean absolute overall score difference was 0.1530 and the mean absolute overall score difference in Cycle 2 was 0.0827.[6]

Given its responsibility for managing the overall program, the NSGO should have greater say when disagreements occur between opinions developed by the PAT over the span of a few days and opinions developed by the NSGO over several years. Conversely, the independent perspective provided by the PAT should be useful to the NSGO when determining which action, if any, to take to address poor performance in these areas.

Some have suggested that larger programs fare better in the FE process than smaller ones. Figure 3.1 plots differences in score between the PAT and FE ratings against program funding as a proxy for program size. The distribution between positive and negative differences does not indi-

---

[5]The numerical score derived is calculated from the numeric equivalent of the four possible ratings, 1 being the highest and 4 the lowest, in each of the weighted criteria. Thus, 1.00 is a perfect score and larger numbers represent poorer performance.

[6]The differences in mean score difference and mean absolute score difference between Cycles 1 and 2 are not statistically significant at the 5 percent significance level of a two-tailed hypothesis test.

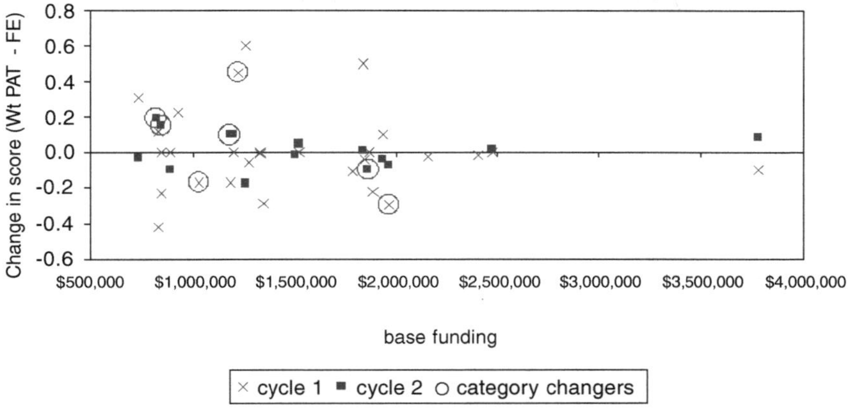

FIGURE 3.1 Base funding (a proxy for program size) vs. difference (change) in overall score during NSGO Final Evaluation review. Category changers (individual program scores that are circled) are the seven programs whose categorization changed (i.e., change in score moved that program either to a higher or lower category within the 5 categories set up by congressional legislation). Four programs improved their categorization and 3 lost ground (data from NSGO).

cate that smaller programs are more likely to receive worse scores in the FE.[7]

Similarly, there has been concern that program officers with long tenure with particular programs might have undue influence in the FE portion of the review (see Figure 3.2). It appears that this concern was unfounded in Cycle 1. In Cycle 2, all programs that received worse scores in the FE (negative PAT-FE) had NSGO program officers with less than 2.5 years with those individual programs. Conversely, all Cycle 2 scores that improved (positive PAT-FE) relative to the PAT score had NSGO program officers with more than 2.5 years with those individual programs. Although all the differences between PAT and FE scores for Cycle 2 were small (< 0.2), two of the changes were statistically significant at the 5 percent significance level.

One of the stated advantages of the FE is that simultaneous consideration of 7 or 8 programs provides an opportunity to compensate for varia-

---

[7]During Cycle 1, the correlation between base funding (2000-2002 average) and changes to the PAT score was -0.108. During Cycle 2, the correlation between base funding (2003) and changes to the PAT score was -0.052. Neither of these correlations is significantly different from zero at a 5 percent significance level of a two-tailed hypothesis test.

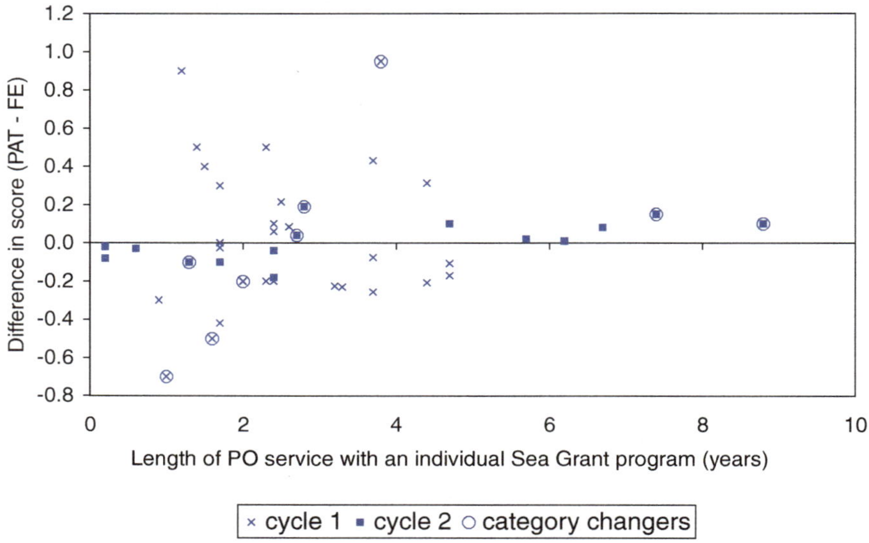

FIGURE 3.2 Continuity of PO service with a particular program vs. difference in overall score assigned by PAT vs. FE. Positive values indicate a better ranking being assigned in the FE (data from NSGO).

tions in the way that different PATs score program performance (i.e., simultaneous consideration of multiple programs helps to address concerns about reliability). To address concerns about reliability and consistency, NSGO staff members would benefit from professional development training in performance evaluation. In addition, an outside expert in performance evaluation could be included in the FE.

IMPROVING THE VALUE OF ASSESSMENT

As noted in Chapter 2, the review process produces a numerical score that is used in allocating merit and bonus funds. Based on testimony and evaluations by committee members expert in this field there is consensus that the criteria set forth as the basis of the review process are appropriate to the goal of improving individual Sea Grant programs. The qualitative ratings for individual criteria are translated into a numerical score and arithmetically weighted to yield a single numerical final score. This section addresses the use of the resulting numerical scores for:

- Determining whether there have been improvements in the individual Sea Grant programs,

- Allocation of merit and bonus funds,
- Identification of potential biases, and
- Broad management of the program.

Improvement

The question of whether the assessment process produces improvement in individual Sea Grant programs can only be judged for the 15 programs that have been through two review cycles. Based on FE scores, the number of individual Sea Grant programs in Category 1 (scores better than 1.5) increased from 7 to 9 between Cycles 1 and 2, the number in Category 2 (1.5 to 2) remained at 5, and the number in Category 3 decreased to 1. Four programs improved their categorization and 2 lost ground. The average ranking number over the entire 15 improved only slightly—from 1.55 to 1.49. Although there was not great improvement, the fact that nearly half of the programs were already in the highest category (scores of 1 to 1.5) implies that there was not much latitude for a major numerical change. In addition, given changes in criteria and benchmarks made during and between cycles, it's not apparent that such relatively small changes in score reflect actual changes in program performance.

The multivariate regression analysis, described in Appendix F, included a variable to reflect differences in average FE scores between Cycle 1 and Cycle 2 while controlling for the influence of other explanatory variables. The results of that analysis suggest that the average difference in scores between the two cycles is not significantly different from zero. Thus, there was no statistical improvement in average program score following the implementation of changes specified in the 2002 reauthorization (P.L. 107–299).

Because the majority of the individual Sea Grant programs receive scores in the "Highest Performance" and "Exceeds Benchmark" categories (categories 1 and 2, respectively), it seems appropriate to wonder if the benchmarks are sufficiently ambitious. If the benchmarks are designed to reflect annually updated, quantitative measures of the significance and impact of research, outreach, and education activities, it would be easier to contrast program performance relative to other programs and to the program's past performance.

The criterion with the most variable results from Cycle 1 to Cycle 2 was "Effective and Aggressive Long Range Planning," with six Sea Grant programs improved and seven downgraded—not a clear indication that the first round led to significant learning. This apparent lack of program change, i.e., the adoption of effective long-range plans, may be remedied if NSGO takes steps, as recommended here, to work with individual Sea

Grant programs to develop and adopt strategic plans. The NSGO should work with the individual programs to generate an agreed upon strategic plan (recommendations for this report can be found in the last sections of chapters 3, 4, and 5). Adoption and implementation of a strategic plan by the NSGO and the individual program would remove the need for a benchmark for the plan itself—establishing the plan would be a joint responsibility. The plan would then be the standard against which the effectiveness of execution would be judged.

Distribution of Merit and Bonus Funds

The practice of awarding "merit" and "bonus" funds based on performance began in 1998 when the NSGO began to emphasize the importance of the new program review process by providing financial rewards for programs that excelled at performance benchmarks. NSGO created three funding categories into which programs were placed based on the scores achieved by each program through the review process. Programs that were ranked in the two best-performing categories (programs with the lowest scores) in the NSGO's scoring system were awarded additional (on top of base funds) or "merit" funding for the duration of the period until their next review. This basic practice continues to this day, with some refinements. Merit funding was intended to reward program performance (based on criteria), rather than competition among programs. It was intended to stimulate improved performance by individual Sea Grant programs

However, in 2002 Congress mandated creation of five sharply defined categories into which the individual Sea Grant programs were to be placed. Congress required "no less than 5 categories, with each of the 2 best-performing categories containing no more than 25 per cent of the programs" (P.L. 107–299, section 3[b][A][ii]). Some consequences of this mandate, which put programs in competition against each other, are at odds with the natural trend and intent of the original merit funding process which was to encourage improvement in all program scores and thereby ultimately aggregate all programs into one category. The NSGO responded to the mandate by retaining the three existing categories and subdividing Category 1 (programs of the highest rank) into three sections (1a, 1b, 1c; scores range 1.0–1.5, 1.5–2.0, and 2.0–2.5, respectively), the first and second of which contain just under 25 percent of all programs. The distribution of scores is shown in Figure 3.3, as of the end of the second year of Cycle 2. The best possible score is 1.0.

Although this adheres to the letter of the legislation, the close numerical spacing of adjacent rankings in Category 1 creates two stepwise discontinuities in the bonus assignment process in which a small difference in score (e.g., between 1.17 and 1.19; or 1.26 and 1.29) results in a

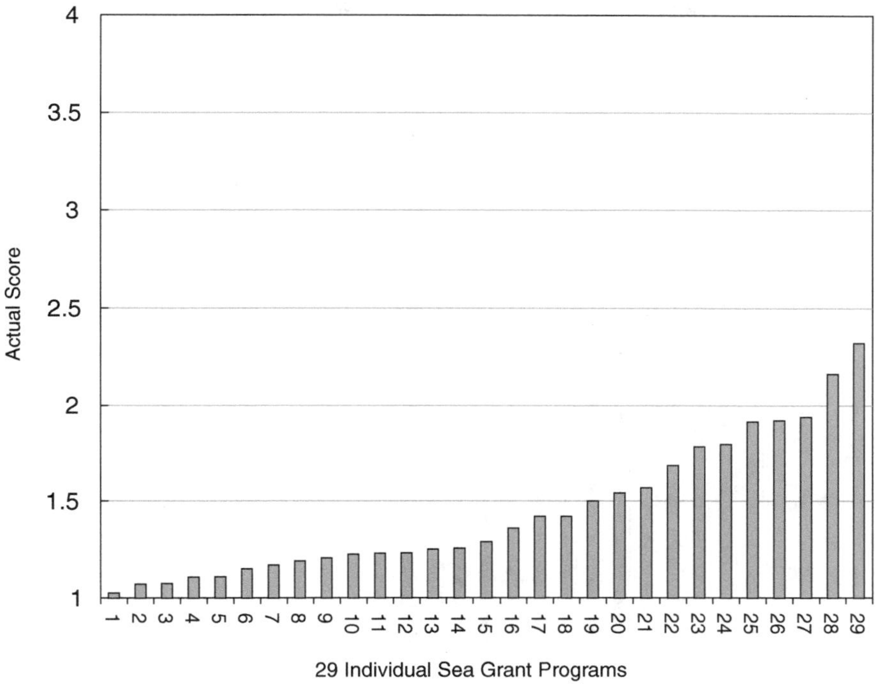

FIGURE 3.3 Distribution of individual program scores in 2005. Each program scored on a 4-point scale with 1.0 being the best possible score and 4.0 being the worst possible score. Merit categories are 1.0–1.5, 1.5–2.0, and 2.0–2.5. Data from NSGO.

significant difference in reward while adjacent corresponding differences have no effect (Figure 3.4). For perspective, the discussion of the FE process notes that the mean magnitude of changes between the PAT and FE scores was 0.1. This small difference in scoring during the FE may have substantial impacts on program funding even though the absolute differences in performance are small.

An alternative to division into discrete categories would be to reward the top 50 percent of the programs on a sliding scale so that there would be no large steps, but rather consecutive small ones. Although there still would be uncertainties in scores at this level of aggregation, a more logical approach would be to reward each program with a bonus increment that would be proportional to the difference in score between adjacent programs. This is the equivalent of computing the bonus in proportion to the difference of any given score in the top half from that of the program at the 50 percent mark in the ranked sequence. The resulting smoothed distribution is shown, based on 2005 data, in Figure 3.4.

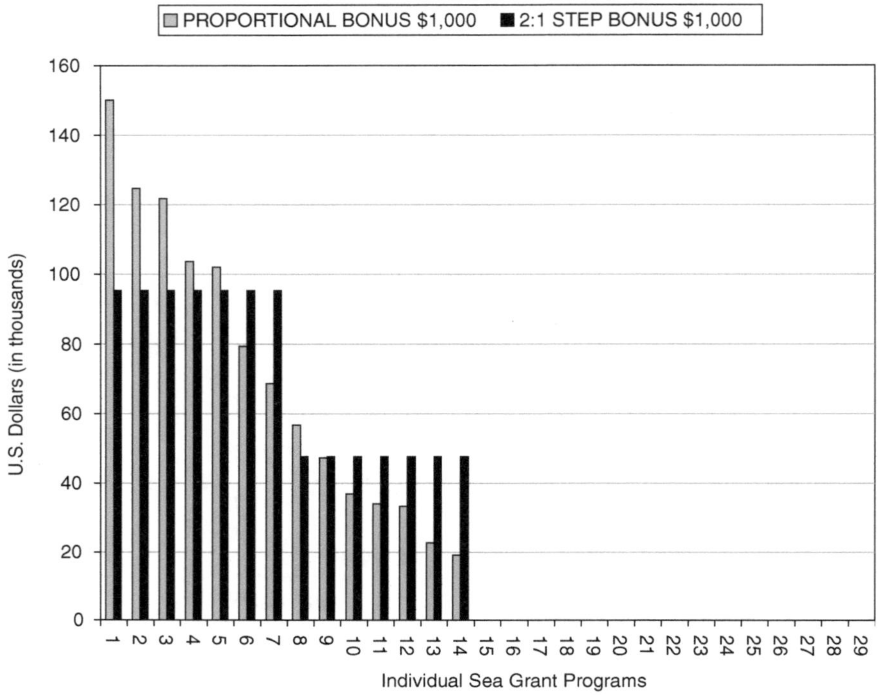

FIGURE 3.4 Current and proposed bonus distributions. Based on the scores of the 29 individual Sea Grant programs, 14 of the 29 programs scored high enough to receive bonus funds from a $1 million pool. Dark gray bars reflect a 2:1 funding ratio between the top seven ranked programs and programs ranked 8 through 14 (note the significant difference between funds awarded to program 7 and 8). Light gray bars show a proposed distribution based on the proportion that each score is above the score of the 15th program. Amounts are in thousands of dollars. Data from NSGO.

Potential Biases

The results of a multivariate analysis of NSGO data, as described in detail in Appendix F, show that the FE scores are not biased as a result of program officer seniority, program funding levels, program maturity, order of review within a cycle, or between Cycles 1 and 2. There is, however, statistically significant evidence that program officer continuity with the individual Sea Grant program is inversely related to the FE score.

Looking at the scores in relation to continuity of NSGO program officer experience with a particular program, Figure 3.5 shows this effect

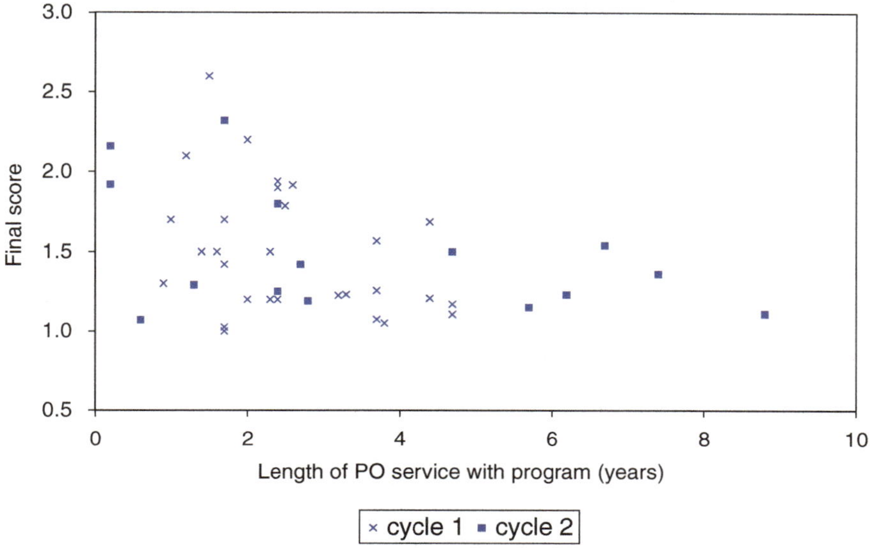

FIGURE 3.5 Continuity of program officer (PO) service with a particular program vs. the FE score. All programs with scores greater than 1.8 are associated with program officers with tenures of less than 3 years. Data from NSGO.

in a simple form: all of the poor scores (greater than 2) occur with NSGO program officers with no more than 2 years of service with that individual Sea Grant program. Considering both cycles, the correlation between NSGO program officer continuity with a particular Sea Grant program and the FE score assigned to that Sea Grant program is -0.37. That is, individual programs that have a longer history of interaction with their NSGO program officer are, on average, assigned lower (better) FE scores.[8]

Two implications can be drawn from these findings. First, the strength of the evaluation system and methodology, in that the final scores are not influenced by variables of the characteristics of the program officer, or the

[8]The probability that a correlation of –0.37 would have been observed if the true correlation is greater than or equal to zero is 0.0066. The coefficients reported in Appendix F are ordinary least squares regression coefficients, not simple correlation coefficients. The multivariate model apportions the observed variation in scores across several different variables simultaneously and thus does not map back to the simple correlation coefficient. In contrast, the simple regression in Appendix F does map back to the simple correlation coefficient: –0.077 = –0.371(1.901)/(0.393), where 1.901 is the standard deviation of program officer (PO) continuity and 0.393 is the standard deviation of the FE scores.

characteristics of the program itself, its funding, and its maturity. Second, while there is evidence of a positive relationship between program officer continuity and the FE score, this relationship should not necessarily be viewed as a cause-and-effect relationship; instead it could be suggestive of the importance of linkages and feedback between the NSGO and the individual Sea Grant programs. The value of robust support by, and interaction with, skilled program officers must be balanced against tendencies for program officers to lose perspective as they develop longstanding relationships with individual SG programs. Rather than serving as a suggestion that scores could be improved by increasing the length of time that NSGO program officers are assigned to particular programs, the statistical finding serves to highlight the importance of ensuring that there is a close and ongoing working relationship between each individual Sea Grant program and the NSGO.

Broad Program Management

Although much of the discussion that took place during open sessions involving individual Sea Grant program directors focused on the use of quantitative scores for competitive ranking of the individual Sea Grant programs, it is important to consider the broader question of the role of the current review process in improving the individual programs and the National Sea Grant College Program (National Program) in other ways. Considerable effort goes into the periodic review process, yet it often appears to be used simply within the narrow confines of assignment of merit and bonus funds. Given the effort involved, the outcomes should be used more widely for program management.

Unfortunately, the dissection of the review into 14 sub-criteria robs the process of an opportunity to take a holistic approach that would enhance its broader application. The PAT and FE discussions become discussions of individual criteria. Roughly as much time was spent in the 2005 FE on a criterion worth 4 percent of the total score as was spent on the research and outreach topics that constitute major contributions (20+ percent) to the total Sea Grant program.

The use of program ratings to rank for competitive funding can have unintended and counterproductive consequences. While competition encourages programs to improve, it can reduce the incentive for individual Sea Grant programs to cooperate with one another or work productively with the NSGO on regional activities. This effect was brought up repeatedly in testimony at public meetings by individual Sea Grant program directors; these directors stated that they were somewhat reluctant to share their ideas with each other for fear of "helping the competition." Sharing and networking have traditionally been important positive elements of the NSGCP and have helped to weave the current 30 individual

Sea Grant programs (not including 3 programs in development) into a single NSGCP.

It is essential that the review process evaluate the manner in which individual programs contribute to the whole. Introduction of an explicit criterion for performance in this area (discussed in the next section) would remedy this shortcoming and improve the effectiveness of the National Program as a whole.

COLLABORATION AMONG INDIVIDUAL SEA GRANT PROGRAMS

In 2004, Admiral James D. Watkins, Chair of the U.S. Commission on Ocean Policy (USCOP), stated in the letter transmitting the Commission's final report *An Ocean Blueprint for the 21st Century* to the President of United States that the USCOP concluded that the following action was essential:

> . . . a new national ocean policy framework must be established to improve federal coordination and effectiveness. An important part of this new framework is strengthening support for state, territorial, tribal, and local efforts to identify and resolve issues at the regional level (USCOP, 2004).

Although the Commission's findings were nonbinding, the heavy emphasis placed on coordination and effectiveness at local, regional, and national scales is striking. Furthermore, in response to the USCOP's report, the emphasis placed on facilitating regional collaboration was adopted in the formal White House response, the *U.S. Ocean Action Plan* (Council on Environmental Quality, 2005). The *U.S. Ocean Action Plan* identified three high-priority actions to address the USCOP's call for "enhancing ocean leadership and coordination." In addition to "codifying the existence of NOAA within the Department of Commerce by passage of an organic act" and "establishing a cabinet-level federal ocean, coastal, and Great Lakes coordinating entity," the Bush administration called for greater effort to support "voluntary regional collaboration." In particular, the *U.S. Ocean Action Plan* underscores support for ". . . enhanced coordination and [the Plan] strongly values the local input that is essential in managing and protecting our nation's ocean, coastal, and Great Lakes resources."

Existing programs, such as Sea Grant, which emphasize local and federal collaboration, would seem to be natural candidates to play leading roles in efforts to address well-recognized and emerging marine policy challenges at regional scales. If Sea Grant can demonstrate an ability to foster regional collaboration, one would expect that ability to be recognized and utilized.

Although some Sea Grant programs are already collaborating at various scales to address issues of high regional interest (such as the Chesapeake Bay area), it appears that these collaborations are driven largely by regional constituencies that interact with multiple Sea Grant programs. Thus it is not apparent that sufficient attention is given in the current review process to systematically identifying opportunities for regional collaboration.

Furthermore, during open session discussions with individual Sea Grant program directors, the assertion was made that the newly enacted Congressional directive to rate and rank programs for the purpose of distributing merit and bonus funds had, to some degree, a chilling effect on program-to-program collaboration. While the veracity of this assertion is difficult to determine, there is reason to believe that the requirement to rate and rank programs has strained the relationship between the individual programs and the NSGO itself. Collaboration is an essential part of integrating the individual Sea Grant programs into a successful National Program. Barriers to effective communication and collaboration among the individual programs could realistically reduce the impact from advances made in various parts of the overall network. Because network building is an important function, it might be advisable to augment the original four criteria with a fifth criterion that assesses the extent to which an individual Sea Grant program contributes to network cohesiveness. Including this additional criterion would ensure that activities in support of the overall network are evaluated in the review process; however, it would only provide insight into one component of the network (i.e., how individual programs contribute to the overall program). In an effort to develop a fuller understanding of how the network is functioning as a whole, greater attention should also be focused on determining how well the NSGO is fostering collaboration at a variety of scales, including supporting collaborative efforts of individual programs.

FINDINGS AND RECOMMENDATIONS REGARDING THE PERIODIC ASSESSMENT PROCESS

The majority of the individual Sea Grant programs receive scores in the "Highest Performance" and "Exceeds Benchmark" categories, thus, it seems appropriate to wonder if the benchmarks are sufficiently ambitious. **The Director of the National Sea Grant College Program, working with the National Sea Grant Review Panel, should carefully review the present benchmarks and indicators to ensure that they are sufficiently ambitious and reflect characteristics deemed of high priority for the program as a whole.**

The evaluation criteria currently used do not adequately emphasize the importance of network building among individual programs and how such activities help to link the local and regional efforts into an effective nationwide program. Some aspects of the current program evaluation process and ranking appear to have fostered an increase in competition and lowered the level of cooperation between individual Sea Grant programs. This tendency is not consistent with efforts to build a cooperative nationwide effort, as encouraged by NOAA guidance documents (33 U.S.C. 1123).[9] Explicit consideration of cooperative and collaborative activities between programs should be included in the program evaluation process and programs should be rewarded for these kinds of activities. Concomitantly, there is no evidence that the use of 14 weighted sub-criteria in Cycle 2 in place of the 4 criteria in Cycle 1 has improved the review process. Conversely, introduction of criteria weighted in small percentages (less than 5 percent) work against taking a holistic view of the individual programs and creates a less efficient process. **The Director of the National Sea Grant College Program, under supervision of the Secretary of Commerce and in consultation with National Sea Grant Review Panel and the individual Sea Grant programs, should substantially reduce the overall number of scored sub-criteria by combining various existing criteria, while adding cooperative, network-building activities as an explicitly evaluated, highly valued criterion.** Benchmarks and indicators for this network-building criterion will need to be carefully constructed so that geographically isolated programs are not inappropriately penalized. However, the steps taken to make such allowances should not undermine the importance of this criterion for the vast majority of individual Sea Grant programs.

Steps taken by the NSGO and the NSGRP to improve consistency in grading are laudable; while it is not possible to attain perfect reliability in a system that values and depends on professional judgments, further actions could be taken to generate improvements in this area. **The Director of the National Sea Grant College Program, working with the National Sea Grant Review Panel, should engage independent expertise to refine the benchmarks and grading instructions to meet professional methods and standards for reliability and to refine the training materials used to prepare individuals involved in the evaluation process, in a manner consistent with the recommendations made in this report.**

While the PAT site visit is a central element of the periodic review, it appears that it has in some instances expanded unnecessarily in terms of

[9]Title 33, Section 1123 of the U.S. Code states that directors shall "encourage and promote coordination and cooperation between the research, education and outreach programs of the administration and those of academic institutions." See Appendix H.

time and cost. Reducing the duration of the site visits would decrease the expenditure of time and funds and allow more overlap of reviewers with increased reliability of the results. Lacking some standards set by NSGO, there is a tendency for individual Sea Grant program directors to expand their presentations to match those of other programs. **National Sea Grant Office and National Sea Grant Review Panel should reduce the effort and costs required to prepare for and conduct a Program Assessment Team site review by providing specific limits on the amount and kind of preparatory material to be provided to the Program Assessment Team and by limiting the site visit to no more than three days, including the time to draft the preliminary report and meet with program directors and institutional representatives.**

The perceived lack of transparency in the FE process has been mitigated by issuance of the 2005 version of the NSGO memorandum describing this phase of the review process. However, lack of transmission of the FE ratings, in contrast to the PAT reports, contributes to a remaining lack of transparency in the FE rating and eliminates a useful opportunity for the NSGO to explain to the individual programs why the views of the NSGO (as reflected in the FE) and the PAT differ.

The "Revised Policy Memorandum on NSGO Final Evaluation and Merit Funding" (NSGO, 2005b) from the NSGCP director moves significantly toward the goal of improving the transparency of these processes (see Appendix E). A few shortcomings remain, particularly the lack of description of how the qualitative ratings of the FE are converted into numerical values and how the merit categories and relative standings are defined in terms of the resulting numerical scores. Greater clarity is needed in the communication of ratings and rankings of programs. **The Director of the National Sea Grant College Program should communicate the results of the FE (annual NSGO Final Evaluation) directly to individual Sea Grant program directors. This communication should include the final rating score received by that program (as begun in 2004) and document any substantial difference between the conclusions reached during the annual evaluation and the most recent periodic review. Furthermore, the Director of the National Sea Grant College Program should communicate the implication of the annual evaluation in terms of the rating and ranking process used to determine a program's eligibility or receipt of merit or bonus funding.**

The diverse score changes for "long-range planning" for the programs that have been reviewed twice show that the long-range planning concept has not been well defined and communicated by NSGO or well implemented by the individual Sea Grant programs. Existence of an appropriate long-range plan shortly after a program is reviewed is essential as a road map for the subsequent interval and as a yardstick against which a program can be measured each year and at the forthcoming PAT review. **The National Sea Grant Office, in consultation with the National Sea Grant Review Panel and individual Sea Grant programs, should establish regular procedures (separate from annual and periodic performance evaluation) for working with the individual Sea Grant program to create and adopt an appropriately ambitious strategic plan, with goals and objectives against which the program would be evaluated at the next program evaluation period.**

There are scoring uncertainties arising from the diversity of programs being reviewed and the differences in interpretation of benchmarks by different PATs such that the stepwise score changes at the 25 percent and 50 percent marks are not defined adequately to justify the abrupt bonus changes at those boundaries. For example, in 2004, 15 programs received bonus funds. An alternative to distributing the bonus funds based simply on whether the program falls into one of only two bins made up the top fifty percent of the programs (by rank) would be to reward the top 50 percent on a sliding scale so that instead of large steps in the award of bonus funds, there would be a gradation of awards. This would reduce the potential for very small differences in scores being converted into large differences in the amount of bonus awarded. **The Director of the National Sea Grant College Program, under supervision of the Secretary of Commerce, should revise the calculation of bonus funding allocation relative to program rank to ensure that small differences in program rank do not result in large differences in bonus funding, while preserving or even enhancing the ability to competitively award bonus funds as required by the National Sea Grant College Act Amendments of 2002 (P.L. 107–299).** Several approaches for accomplishing this seem worthy of consideration. One approach would be to reward each program in the upper half in proportion to the difference between its score and the score of the program at the 50 percent mark (the median score). The resulting smoothed distribution is shown in Figure 3.4. Another possible alternative would be to smooth the distribution based on the relative standings of those programs in the top half relative to the middle program. This second approach is less attractive given that the relative standings are themselves derived from the program scores. Neither approach would totally eliminate differences in bonus funding between programs that have statistically similar scores, but either approach would significantly reduce the potential for two programs with statistically similar scores from receiving significantly different bonus awards. Both approaches would appear to satisfy the congressional desire to see bonus funding distributed based on performance (P.L. 107–299).

RETHINKING THE PROGRAM ASSESSMENT PROCESS

Many of the changes proposed above are intended to address the challenges to effective program assessment that stem from the desire to rate and rank the individual Sea Grant programs for the purposes of determining which programs qualify for bonus funding and to support efforts to distribute funds in a competitive manner. As discussed in Chapter 2, in response to congressional desire to see a greater level of oversight and competition in the program, the purpose of assessment within the Sea Grant program became two-fold. First, and more traditionally, assessment is used to identify weaknesses or opportunities for growth in the individual Sea Grant programs and possible mechanisms to address them. Second, and more recently, assessment is used to reward programs for achievement (i.e., rate and rank programs in order to pass out bonus funds competitively).

Steps proposed to further strengthen the assessment process for the purposes of establishing a more credible and reliable rating and ranking system (including greater overlap among PAT teams, more uniform PAT visits and briefing materials, shortened PAT visits to allow completion of the PAT reviews in a shorter period, etc.) may be difficult to fully achieve and, would likely reduce the value of assessment for the purpose of exploring areas of growth or mechanisms for accomplishing it. Thus, it would seem appropriate to explore an alternative structure for assessment within the Sea Grant program, one that fundamentally embraces the two purposes of assessment, by developing two separate mechanisms, each tailored to address a single, more or less unique purpose.

Designing an effective dual-mode assessment process would require that one mode emphasize the main purpose supporting the annual rate and rank process, while the main purpose of the second mode would be to nurture the program by evaluating the National Program in its entirety (i.e., all the individual programs as well as the NSGO) at least once every 4 years.

Such a change in approach would allow external peer reviewers to move beyond simple ratings to consider broader issues such as an independent check on individual programs and the evaluation process overall. Broader issues may include identifying areas for growth or improvement and mechanisms for achieving that growth or improvement, exploring ways to strengthen the individual programs institutional relationships, examining the nature of the individual program's relationship with the NSGO, and the effectiveness and credibility of annual evaluation (to support findings about the "state" of the individual programs as well as the network overall). The implications of such a change will be further explored in Chapter 4.

4

Program Oversight and Management

In the United States and around the world, research programs are funded through a variety of mechanisms and use different project selection and program evaluation approaches to maintain quality of performance. However, the trend across research programs worldwide is toward more competitive project funding and stronger retrospective evaluation processes, with stronger links to resource allocation. In keeping with this trend and in compliance with congressional directive, the National Sea Grant College Program (NSGCP) funds are distributed through: (1) centrally managed competitive awards to investigators, (2) awards based on historical factors of individual Sea Grant programs and subsequently distributed as competitive awards to investigators, and (3) competitive bonus awards to individual Sea Grant programs based on their relative ranking as a result of the program review process.

Although the program oversight, structure and management processes followed by the NSGCP are somewhat unique, program review by other federal, state and private grant programs share similarities and dissimilarities with Sea Grant's review program. This chapter will discuss these similarities and dissimilarities, especially in regards to the six main elements of administration used by the Sea Grant program, (introduced in Chapter 2): (1) annual reports prepared by the individual Sea Grant programs; (2) sporadic interactions with the National Sea Grant Office (NSGO) administrators and program officers; (3) periodic assessments by high-level external review teams (Program Assessment Teams [PATs]); (4) certification reviews for aspirant and deficient programs; (5) the development, approval, and implementation of strategic plans at the national and individual levels; and, (6) annual allocation of federal funds (base, merit, bonus, national initiative, special projects).

INTRODUCTION TO PROGRAM MANAGEMENT

Program management and program oversight are interrelated processes that directly affect the success of a program in achieving its mission, goals, and objectives as described in the strategic plan. Effective program management requires acquisition of information about program performance, outcomes, and impacts through a combination of continuous and periodic processes. Program administrators use this and other information to evaluate progress, allocate resources, and make decisions that influence the direction and focus of the program in implementing the strategic plan. Effective program oversight includes both internal mechanisms and external mechanisms at a variety of time scales. Internal monitoring and oversight occurs on continuous or short time scales (days, weeks, months) to drive short-term decision-making at a local program level, while external monitoring and reviews take place on longer time scales (semi-annual, annual, quadrennial, etc.) to inform national program decisions and long-term local program decisions. A key element of effective management and oversight in any program, especially a dispersed national program such as Sea Grant, is a strategic plan that is integrated throughout the program structure, bringing cohesion to the effort without eliminating the focus on local challenges, opportunities, and networks.

MANAGEMENT AND OVERSIGHT OF RESEARCH AND OUTREACH PROGRAMS

The Sea Grant program is certainly unique within the National Oceanic and Atmospheric Administration (NOAA), and perhaps within the federal government as a whole, in terms of the use of annual rating and rankings to determine eligibility for and size of bonus funding. However, the use of formal performance reviews of individual programs is common. Federally supported research, outreach, and education programs are funded through a variety of mechanisms, including broadly competed awards, formula-funded block grants, and funds budgeted to and expended within federal agencies. In addition, funds can be awarded for short or long periods to individual investigators, small teams, centers and institutes, or national labs. In this section, a handful of different programs are described to provide examples of existing federal programs and show the diversity of program oversight and assessment processes.

Sea Grant is one of a handful of federal research programs that provide block funding, on either a fixed or formula basis, to universities for research, outreach, and education. The largest such federal program and the conceptual model for Sea Grant, the U.S. Department of Agriculture administered Cooperative State Research, Education and Extension Service (USDA-CSREES), uses formulae to allocate about $550 million annually to public land grant colleges and universities pursuant to several federal laws[1] (See Box 4.1). While some of the USDA-CSREES formula-funded programs are of recent origin, new funding for research and outreach has been increasingly directed to competitive programs. In addition, USDA-CSREES has come under increasing pressure to shift funding from formula programs to competitive programs.[2] For example, the Annual CRIS (Current Research Information System) report, a different report from the follow-up report, is required and is prepared by principal investigators. The CRIS report includes descriptions of each research project, project outcomes, and impacts. The CRIS database can be queried to parse reports by geographic region, subject of investigation, individual investigator, etc. The annual reports are used by the State Experiment Station directors to oversee and assess the productivity of funded projects. The initial FY06 budget proposal included a $250 million increase for National Research Initiative (NRI) competitive grants, funded in part from a $104 million reduction in formula funds.

The CREES reviews have this flexibility because the need for intercomparison among reviewed programs is less and the programs are not ranked for the purpose of making funding decisions.

[1]Hatch Act; Smith-Lever: 1862 Institution; Smith-Lever Act 3(d); Food and Agriculture Defense Initiative; Renewable Resources Extension Act; McIntire-Stennis Cooperative Forestry; Animal Health and Disease Formula; Aquaculture Centers; Evans-Allen 1890 Research Formula; 1890 Extension Formula; 1890 Facilities Grants; 1890 Institutions Teaching and Research Capacity Building Grants; Tribal Colleges Endowment Fund; Tribal Colleges Education Equity Grants; Extension Services at the 1994 Institutions; Tribal Colleges Research Grants; Hispanic Serving Institutions Education Grants; Resident Instruction for Insular Areas; Alaska-Native Serving and Native-Hawaiian Serving Institutions Education Grants; and Agriculture in the Classroom.

[2]Some examples of comparable programs: National Research Initiative; Sustainable Agriculture Research and Extension; Outreach and Assistance for Socially Disadvantaged Farmers and Ranchers; Organic Agriculture Research and Extension Initiative; Higher Education Challenge Grants; Secondary and Two-Year Postsecondary Agriculture Education Challenge Grants; Food and Agricultural Sciences National Needs Graduate and Postgraduate Fellowship Grants; Multicultural Scholars; International Science and Education Competitive Grants; Biotechnology Risk Assessment Research; Small Business Innovation Research; Community Food Projects; and Risk Management Education.

Box 4.1
Case Study: Program Review Processes in a Formula-Based Block-Funded Program: U.S. Department of Agriculture-Cooperative State Research, Education, and Extension Service (USDA-CSREES)

The periodic review and evaluation process of CSREES programs is similar to the Sea Grant review process but with significant differences. In both programs an outside panel of experts is convened to review pertinent program documents, conduct a site visit to the program, and report findings and recommendations based on a set of pre-determined review criteria. However, unlike the Sea Grant program reviews, which are previously scheduled, the local program administrator requests a CSREES program review 18 to 24 months in advance of the review, often in conjunction with a university review. The purposes of a CSREES review are specifically to "assess the benefits that the programs provide to the agricultural industry, rural communities/environment, and consumers, and in meeting other social goals stated in congressional authorizing legislation" (CSREES, 1999). While most reviews comprehensively address research, outreach, and instructional programs, some are designed around specific issues or programs.

Unlike a Sea Grant program review in which the scope and timing have been determined by the NSGO and expressed in the PAT Manual, the scope, purpose, and timing of a CSREES review are determined through consultations among the Experiment Station/Extension administrator, program leader, and CSREES team leader. Specific objectives for the review, a general timeframe for the site visit, and the size of the review team and specific areas of needed team expertise are all determined through these consultations and all parties reach a verbal agreement.

Composition of CSREES review teams is similar to the composition of Sea Grant PATs although selected through a different process. Review teams include 4 or 5 members, selected by the CSREES team leader based, in part, on nominations by the local program administrator and program leader. The team usually includes one member as an institutional representative—a department head or program leader from a related department—of the program being reviewed. Other team members are selected for their recognized knowledge and experience relevant to the review objectives and usually include department heads, program leaders, or Experiment Station/Extension administrators from peer institutions.

The review team approves the site visit agenda four to five months prior to the review. Site visits are usually two to three days, including time for deliberation and writing the report. Reviews may include visits to laboratory and other facilities, slide and video presentations, structured meetings with faculty and administrators, and unstructured interactions with faculty, students, and staff.

At the end of the site visit, the review team presents preliminary findings and recommendations to campus administrators and to the faculty. The review team has four weeks to submit a final written report to the CSREES administrator, who then writes and forwards a final report to the Experiment Station/Extension administrator within an additional two weeks. CSREES requests that the program leader/department head or Experiment Station/Extension administrator submit a follow-up report about one year after the review outlining actions taken in response to the review. However, since this follow-up report is not required, it is rarely completed.

The Sea Grant Review Process Compared to Other Federal Programs

The stability of base funding for individual Sea Grant programs is like the funding stability enjoyed by government laboratories. In some government laboratories, such as the intramural laboratories of the National Institutes of Health (NIH) and the Agricultural Research Service (ARS). NIH and ARS researchers are federal employees, so personnel evaluation procedures help to maintain quality. But such laboratories typically also have strong program review processes. NIH laboratories, for example, are reviewed by external teams every 4 years (M. Gershengorn, NIH, personal communication, 2005), and ARS laboratories every 5 years (www.ars.usda.gov). ARS funds activities through proposals that are externally reviewed and that must meet minimum quality requirements. The national laboratories funded by the Department of Energy and managed by contractors also review all projects before they are approved for funding.

The individual Sea Grant programs resemble some additional aspects of federally funded centers. For example, NIH, National Science Foundation (NSF), and other U.S. federal agencies, provide multiyear awards to centers that perform multiple functions, usually including research; education; outreach; in some cases, service (e.g., in NIH centers, translation into clinical practice); and the provision of central infrastructure. These awards are usually made for four or five years at a time. At approximately mid-award cycle, NSF centers receive a site visit. Performance is closely examined in relation to milestones specified in the original proposal. At both NSF and NIH, centers must submit applications for continuing funding every five years. These requests for continued funding are considered in competition with other projects considered for funding. At NSF, most programs have sunset clauses that limit funding for individual centers to a maximum of ten years, or two five-year awards. Centers are expected to be self-sufficient after that time.

Implications of Review on Funding and Competition

The individual Sea Grant programs could be described as centers funded by the NSGCP. They differ from NIH and NSF centers in that their locations and base budgets were not openly competed, the base funding is not subject to regular recompetition, and they are not subject to a sunset clause unless their performance warrants decertification. Individual Sea Grant programs also differ from NIH and NSF centers to the extent that nonfederal funds, especially state funds,[3] are used for their support.

[3]The average reported share of state funding across all the state programs from 1995–2003 is 35.2 percent. Because many individual programs only report to the NSGO those nonfederal funds needed to demonstrate the match to federal funds, this amount is likely the minimum.

The initial certification of an individual Sea Grant program does not involve a national request for proposals (RFPs) and open competition among perspective institutions, but instead involves a review process wherein the candidate program prepares a proposal that describes institutional and state support, program leadership and organization, an analysis of stakeholder needs, and the identification of priority research, outreach, and education programs. Programs that fail to demonstrate excellent potential are not certified. Further program evaluation may also occur in the course of annual program review if there is a persistent failure to address shortcomings identified in the periodic review or if there is a loss of critical personnel or institutional support. To date, no individual Sea Grant College program has been decertified as a result of poor performance.

The principal alternatives to block funding of research, education, and outreach programs are (1) the funding of in-house research by agency or organization staff and (2) project funding through peer review of competing proposals. Most research grant programs supported by NSF and NIH, and other federal research programs (such as the U.S. Environmental Protection Agency Science to Achieve Results [EPA-STAR] and USDA-NRI grant competitions) are characterized by a broadly distributed open public solicitation for proposals, peer or panel review of the proposals and credentials of the principle investigators, and administrative review of the highest rated proposals and determination of funding levels. The grants may be awarded to individuals, teams, or research centers. NSF, NIH, and USDA-NRI typically fund 20 to 30 percent of the proposals submitted. The EPA-STAR program funds about 15 percent of the proposals submitted. For a concise description of the oversight and management processes of these and related federally funded research programs, see NRC (2001). Although Sea Grant sponsors some nationwide open competitions for research funding, most Sea Grant research funds are allocated as formula-based block grants to the individual Sea Grant programs, where funds are allocated to outreach and education programs, program administration, and through competitive awards to investigators.

Allocation of Funds, Peer Review, Competition, and Awards to Meritorious Projects

In the case of individual Sea Grant programs and USDA-CSREES formula-funded programs, there is a second-stage allocation of funds at a state level, which usually relies on peer review of competing proposals. For example, individual Sea Grant programs hold biannual competitions for research funding. While the specific details of the competition vary across programs, all of the competitions include anonymous peer review of the scientific merit of the project and the qualifications of the principal investigator (PI) or investigators. Many include review by a panel of stakeholders to score the proposals for their relevance to critical regional needs and all include a final technical review by a panel of peers supervised by the program director. Projects to be included in the program's omnibus proposal to the NSGCP are selected during the final panel review.

In the case of USDA-CSREES, the formula funds allocated to state agriculture experiment stations include a second stage-allocation that typically involves peer evaluation of research proposals, but the peer evaluation is usually in-house (at the institution) and is designed to provide advice about improving projects rather than to screen and identify projects to be funded. Within the land grant system, USDA-CSREES formula funds have been built into the base funding for tenured and tenure-track faculty and so it is problematic if state experiment station directors deny salary funding for projects, and less problematic if they deny operating funds. Because operating funds are typically small relative to salary funds, the state experiment station directors have limited ability to promote strong projects or eliminate weak projects.

In contrast, because Sea Grant projects have not been captured into the base salaries of research faculty, individual Sea Grant program directors have more flexibility to shift funds to meritorious projects. Although the competition for Sea Grant research funds is not strictly a national competition, the competition at the individual Sea Grant program level is intense (less than 20 percent of the proposals are funded and many proposals are funded at less than the requested level). Moreover, proposals are broadly solicited and often involve PIs and co-PIs who are not associated with the university or consortium that hosts the individual program. While there is within-program competition for research funds in the formula-funded USDA-CSREES and individual Sea Grant programs, there has not been, but could be, a comparable within-program competition for outreach and education funds.

STRATEGIC PLANNING AS A PROGRAM DEVELOPMENT AND EVALUATION FRAMEWORK

Strategic planning is a cornerstone of effective program management. A well-designed strategic plan reflects goals and objectives that the program intends to accomplish within the planning horizon. In a disaggregated and regionally dispersed program, strategic planning could help integrate individual programs into a national whole while supporting regional and local differences in program emphasis. Weaknesses in the strategic planning process and the lack of effective integration of local and national strategic plans were recognized by Duce et al. (2002). The NSGO has responded to some of the recommendations of Duce et al. (2002) and to the increased emphasis that the federal Office of Management and Budget (OMB) and the National Oceanic and Atmospheric Administration (NOAA) have placed on strategic planning (NSGO, 2004b). Similarly, some individual Sea Grant programs have developed strategic plans that reflect active collaboration with the NSGO as well as its local constituents. However, other individual Sea Grant programs have been slow to develop strategic plans or have strategic plans that are poorly designed, poorly integrated with the national strategic plan, or lack specificity for addressing local and regional needs.

As noted in NRC (1994) and Duce et al. (2002), information from the individual Sea Grant programs is extremely valuable in the development of national priorities and objectives. The individual Sea Grant programs have direct contact with researchers, educators, outreach specialists, and stakeholders in the marine community and are thus well positioned to identify emerging issues. While the goals and objectives expressed in each program's strategic plan can be expected to address issues of uniquely local importance, it is essential that they also be placed in the context of the NSGCP strategic plan, which is modified every four years to comply with U.S. Code. Conversely, the formation of a cohesive integrated national program with discernible regional goals that could be addressed through the combined efforts of individual Sea Grant programs, would require that the strategic plans for each individual Sea Grant program include elements that are common with the national plan as well as elements unique to the locale, including those elements that address needs identified by the states and other sources of financial support. Hence, there is a need for top-down and bottom-up integration of strategic plans. While integration is important for overall program coordination and oversight, the NSGCP strategic plan should be more than a simple collation of the strategic plans developed by the individual Sea Grant programs, and the individual strategic plans should be more than a simple subset of the NSGCP strategic plan. Development of strategic plans for individual Sea Grant programs presents a prime opportunity to strengthen interactions with the NSGO and regional or thematically relevant sister programs.

An effective integrated strategic planning process could begin with the development of an appropriately ambitious draft strategic plan with input from key stakeholders, university or consortium administration, and the NSGO. When formally approved by the National Director, the individual Sea Grant program's strategic plan represents a compact between the individual program and the network as a whole. Approval by the National Director signifies that the program's strategic plan is sufficiently ambitious and attendant to local, regional, and national priorities, so that successful and timely accomplishment of the goals and objectives outlined in the plan can be expected to result in superior or outstanding ratings for corresponding elements of the annual and periodic program reviews. In turn, when an approved strategic plan is in place, annual reports and periodic program reviews can be framed in the context of accomplishments relative to goals and objectives outlined in the strategic plans in effect during the review period. Programs that achieve the identified goals should be assured of receiving superior or outstanding ratings.

Because the NSGCP is required to prepare a new strategic plan every four years, there are advantages to having the individual Sea Grant programs prepare or update their strategic plans on a coincident cycle. Harmonizing the periodicity of the strategic planning process and the periodicity of program review would allow the program review process to look back at performance relative to strategic plans in place during the review period. The program review process could also look forward to the strategic plan that has been developed for future activities and could comment on the significance of the activities proposed and the availability of resources to support those activities.

ROLE OF THE NATIONAL SEA GRANT OFFICE

The role of the NSGO was examined recently by Duce et al. (2002). A comprehensive reexamination of the NSGO is outside the scope of this review, but an evaluation of the Sea Grant program review process cannot be entirely decoupled from consideration of the role of the NSGO in that review process. Effective program administration within a diverse and decentralized organization such as Sea Grant requires a clear and consistent process for providing the central organization with accurate and comparable information about the objectives, activities, and performance relative to those objectives of the decentralized elements of the organization. In addition, there must be a clear and consistent vehicle for conveying information about current and anticipated goals and objectives from the center of the organization back to the individual programs. The National Director, working through the NSGO, is responsible for ensuring that there are effective conduits of top-down and bottom-up information flows. However, based on discussions with individual Sea Grant program directors and with NSGO administrators and program officers, it is evident that NSGO personnel have limited interaction with the individual Sea Grant program directors and that top-down and bottom-up information conduits are less than effective.

NSF and many other federally funded research programs rely on program officers for ongoing communication between distributed programs and national program administrators. To be effective as the primary top-down and bottom-up information conduits, program officers must receive training in program evaluation and administration, and must have backgrounds in the technical disciplines of the programs with which they interact. Irrespective of whether NSGO relies on program officers to serve as the primary information conduit, it is essential that some structure be in place to serve this function. Effective program evaluation depends on the degree to which assessment is normalized by the national office, based on the objectives and program planning of individual programs rather than some preconceived standard. Because of differences among individual programs (financial resources, talent pools with various specialties, issues, approaches, geographic and demographic characteristics) and the unique institutional environments, assessment should be tailored to take into account program variability. The NSGO program officer could be the link between the NSGCP and the individual program directors, providing the perspective for assessing program effectiveness annually while considering institutional characteristics.

ANNUAL AND PERIODIC ASSESSMENT PROCESSES AS INTEGRAL ELEMENTS OF PROGRAM ADMINISTRATION

Periodic program assessment within the NSGCP is intended to serve two related purposes. The first, narrower purpose, is to fulfill the congressional mandate to rank programs for the competitive award of merit and bonus funds. The second purpose is to identify areas for improvement in individual programs. These are related insofar as competition for funds serves as an incentive to the individual programs to improve and the periodic program assessment process provides information that can be used to direct improvements where necessary. The periodic assessment process, as it has evolved, appears to be aimed disproportionately at the narrow goal of ranking programs and distributing competitive funds. Although this is understandable given the congressional mandate, an assessment process that is excessively geared toward ranking the programs may do a poor job in other aspects of program improvement. For example, while the episodic interactions between the NSGO and the programs may be sufficient for ranking, it may not provide sufficient timely information for directing program improvements. Similarly, the National Sea Grant Review Panel (NSGRP) has become overly concerned with the periodic assessment process. All this implies that simply tinkering with the PAT manual and eliminating discontinuities in the way in which competitive funds are distributed will not solve the problem. For the program to improve—and, in particular, for it to become and be perceived as a truly national program—there is need for more individual Sea Grant program involvement with the NSGO and there is a need for more NSGO involvement with the individual Sea Grant programs. At the same time, the role of the NSGRP in the assessment process needs to evolve. The extent to which the NSGRP has become involved in the details of the periodic assessments is a reflection of the overreliance on these assessments. **The Director of the National Sea Grant College Program, in consultation with the National Sea Grant Review Panel, should work to establish an independent body to carry out the periodic assessments under the supervision of the National Sea Grant Review Panel.** By removing itself from direct involvement in the individual program assessments, the NSGRP will be better positioned to comment on issues of broader significance to the overall program, including efforts by NSGO and the individual Sea Grant directors to strengthen the partnership aspects of the NSGCP. The NSGRP should continue to monitor the process closely, but should be perceived as a neutral body whose sole function is to promote the effectiveness of the program as a whole.

The purpose of program oversight and management is to ensure that the program managers are aware of the array of activities that are being undertaken and to ensure that program managers have a basis for program assessment so that resources can be managed to improve the capacity and performance of program components. Strong program oversight and management systems blend ongoing and annual assessment of program activities and outcomes with periodic assessments that explore the long-term effectiveness of programs and consider the summation of accomplishments, outcomes, and impacts. In addition, a periodic assessment provides external validation of the annual program assessment (Box 4.2 gives an example of another federal program, the Louis Stokes Alliances for Minority Participation [LSAMP] grant program, with periodic assessments and external reviews and illustrates a reverse review concept).

Although NSGCP has annual reporting requirements, ongoing interactions between the NSGO and the individual Sea Grant programs, and a periodic program assessment process, the information provided through the annual reports and ongoing interactions between NSGO and individual programs could play a more prominent role in the annual assessment of programs, and specifically, could provide information for program oversight and management. Ongoing and annual assessments are essential for effective program management. To effectively administer the program, the NSGO must be aware of the activities and accomplishments of and opportunities and challenges faced by the individual Sea Grant programs. The National Director cannot effectively convey information

Box 4.2
Case Study: Program Review Processes in a Competitive Grant Program: Louis Stokes Alliances for Minority Participation

The objective of LSAMP, a program managed by NSF, is to increase the quality and quantity of undergraduate baccalaureate recipients in the natural sciences, mathematics, engineering, and technology, with particular focus on students underrepresented in these areas. Each LSAMP project is administered through a five-year cooperative agreement between the academic institution and the NSF. Although the awards of $2.5 to $5 million are for a full five-year cycle, the cooperative agreements are administered as five one-year contracts, with continuing funding contingent on achieving satisfactory progress as determined by a three-part annual evaluation.

The program review and evaluation process of LSAMP programs shares some commonalities with Sea Grant's evaluation process. In both, an outside panel of experts conducts a site visit and reports its findings and recommendations based on an assessment of the effectiveness of each activity in supporting program goals. However, unlike the Sea Grant program assessments, the site visits for LSAMP programs are conducted annually and represent only one part of the program review process. In addition to the annual site visit, the annual evaluation process for LSAMP programs includes an annual report and a reverse site visit (LSAMP program officers visit NSF). The NSF reviews the results of all three components of the evaluation, *in toto*, before deciding whether the next year of funding should be awarded. This enables the agency to terminate projects that fail to achieve agreed upon goals. The NSF assigns a single program officer to the LSAMP program whose sole responsibilities are the administration of LSAMP projects. The program officer serves as chair of the annual site visit team.

about these activities, accomplishments, opportunities, and challenges to NOAA, DOC, or the Congress unless the information is readily available.

As recommended in Chapter 3, performance metrics are needed that can be readily validated, and that can assess the quality and significance of program activities, outcomes, and impacts. If the annual reports describe activities, outcomes, and impacts in terms of the same metrics that form a basis for the periodic PAT reviews, then the annual ranking of the individual Sea Grant programs could be based on a combination of information submitted in the annual reports, information available to the NSGO through other reporting requirements, and interactions between NSGO representatives and the individual Sea Grant programs, augmented by the PAT reports and the individual program directors' responses to the PAT reports. By viewing the PAT report as only one, albeit important, source of information feeding into program assessment, con-

The LSAMP site visits are short—one to two days—and focus on assessing the effectiveness of new outcomes that would not have occurred in the absence of allocated resources. Similarly to the Sea Grant process, the site visits follow standardized protocols and assess performance relative to well-defined criteria, including degree production and enrollment data, expenditures of NSF and nonfederal matching funds, and programmatic activities. Site visit teams focus on (1) preactivity status; (2) postactivity results; and (3) net changes as they relate to enrollment, retention, and degree production goals. Unlike the Sea Grant PATs, the LSAMP site review teams do not produce a report that reifies performance into a single numeric rating, but instead the review team provides a report that discusses strengths and weaknesses of the program.

The LSAMP annual reports, the second component of the annual review process, follow a template defined by NSF. While these reports provide flexibility for programs to report on a wide variety of activities, the template prescribes inclusion of standardized metrics, the "Minimum Obligatory Set" (MOS), which include key performance indicators. The MOS data are used to track program performance through time and make interproject comparisons. Unlike Sea Grant, project rankings are not a part of the evaluation process.

At the reverse site visits, the third component of the annual evaluation process, each LSAMP program participates in a one-hour review at NSF headquarters with the program officer and other NSF administrators. These (individual) sessions consist of a brief presentation by the LSAMP program team followed by a discussion of data reported in the program's annual report and the findings and recommendations of the site visit team. All reverse site visits are scheduled during a single week, so that all programs can be evaluated within the same period of time.

SOURCE: NSF, 2003.

cerns about the asynchronicity of periodic assessments will be lessened. At the same time, more fully incorporating the annual reports and other ongoing information (communicated via miscellaneous documentation, e-mails, phone conversations, general site visits by the program officer, and program interactions) regarding program activities and progress, will ensure that the annual ranking is based on the most recent information about each individual program.

FINDINGS AND RECOMMENDATIONS REGARDING PROGRAM OVERSIGHT AND MANAGEMENT

As discussed in Chapter 3, the periodic assessment process, as currently carried out during PAT site visits and NSGO Final Evaluation Review (FE), will require some modification to increase its reliability and credibility for the purposes of rating and ranking individual programs in a manner that will support the distribution of merit and bonus funding. The more important issue identified here is the need to fit periodic assessment into the larger effort to continually improve and enhance how periodic assessment should fit into a larger effort to continually improve and enhance the overall program as it strives to provide "an appropriately balanced response to local, regional, and national needs" (33 U.S.C. 1123) is needed.

Sea Grant Program Administration

Since the 1994 report, the NSGCP has significantly improved the structure of its management and oversight processes. The improvements include a stronger strategic planning process, decentralized and professional review of project proposals, and a robust program review process. Altogether, the NSGCP oversight processes include annual allocation of federal funds (base, merit, bonus, national initiative, special projects), periodic reviews of national and individual Sea Grant program processes and outcomes (PAT visits and reports, certification reviews for aspirant and deficient programs), regular monitoring of national and individual Sea Grant programs (annual reporting, interactions with program officers), and the development, approval, and implementation of strategic plans at the national and individual Sea Grant program levels. In practice, however, many elements of NSGCP's program oversight system have atrophied, and program oversight has essentially been reduced to the PAT site visits and the ratings and report that derive from the PAT visits. This overreliance on periodic review of outcomes and impacts fails to provide timely, ongoing feedback to the NSGO throughout the review cycle and diminishes the effectiveness of program oversight. **The Director of the National Sea Grant College Program should ensure that program administration carried out by the National Sea Grant Office makes full and consistent use of annual reporting, frequent and meaningful interactions with individual Sea Grant programs by National Sea Grant Office program officers, and the development, approval, and implementation of strategic plans to monitor and assess the performance of the individual Sea Grant programs on an ongoing basis.** Reverse site visits (see LSAMP case study, Box 4.2) appear to be a viable mechanism for connecting individual Sea Grant program directors with program officers and NSGCP administrators, and would likely provide an opportunity for the National Director to evaluate the nature of the relationships between NSGO staff and the individual Sea Grant programs, and for collective discussion of near-term planning and information exchange. The intent of the reverse site visit suggested here is to ensure that the

NSGO is responsive to its state and local partners; the reverse site visit should not be used as a substitute for NSGO program officer visits to individual Sea Grant programs.

Periodic program assessment is an important external check on the effectiveness of both the individual Sea Grant programs and the NSGO's ability to facilitate and coordinate their efforts. **The Director of the National Sea Grant College Program, working with the National Sea Grant Review Panel, should redirect the focus from periodic external Program Assessment Team reviews towards identifying areas and mechanisms for improving the individual Sea Grant programs as well as the National Sea Grant Office's efforts to facilitate and coordinate program efforts.** External, periodic review can thus provide an independent snapshot of program performance in areas assessed annually by the NSGO. **The Director of the National Sea Grant College Program, in consultation with the National Sea Grant Review Panel, should create a process for determining the underlying causes of disagreement for instances where a Program Assessment Team review appears to reach conclusions at odds with the most recent annual assessment provided by the National Sea Grant Office.**

Role of the National Sea Grant Office

The NSGO does not currently play a sufficient role in ongoing program assistance, monitoring, communication, and assessment, nor does it maintain close ongoing working relationships with the individual Sea Grant programs. There were more interactions and better relationships between the NSGO and the individual Sea Grant programs prior to 1995. As noted in Duce et al. (2002), closer and more frequent interaction with NSGO would help integrate individual Sea Grant programs into the National Program. **In order to effectively administer the Sea Grant program, the Director of the National Sea Grant College Program should take steps to ensure that sufficient qualified staff are available to interact with the individual Sea Grant programs, to ensure effective two-way communication, and to monitor and assess program performance on an ongoing basis.**

Strategic Planning Process

Strategic planning is key to effective management and oversight of the individual Sea Grant programs. Strategic planning is not well integrated into the NSGCP despite the fact that strategic plans are a specific criterion in the program assessment process. Sea Grant program strategic plans do not reflect active collaboration between the NSGO, the individual Sea Grant program, and the institutional representative. Many individual Sea Grant programs have strategic plans but the quality varies widely. Although some programs submit their strategic plans to the NSGO, those plans are neither formally reviewed nor approved by the NSGO except as part of the PAT. **Each individual Sea Grant program, in collaboration with its local network and the National Sea Grant Office, should develop an appropriately ambitious, high quality strategic plan that meets local and institutional needs while simultaneously reflecting the individual program's role in addressing the regional and national needs identified in the strategic plans of NOAA and National Sea Grant College Program.** The plan should include clearly articulated goals, tailored to the individual program that can form the basis of annual and periodic performance evaluation. In other words, the benchmarks of performance in each area should be jointly developed by the NSGO and the individual Sea Grant program, and incorporated into the strategic plan of each program, through a process separate from either the annual or periodic performance evaluation. Coordination between the individual Sea Grant program director and the NSGO on strategic planning can also provide the NSGO with feedback on local trends and shifts in local and regional perspectives, which could improve the content of future NSGCP strategic plans. **The Director of the National Sea Grant College Program, in consultation with National Sea Grant Review Panel, should formally review and approve each individual strategic plan.** The approved strategic plan would then serve as the basis for annual and periodic evaluation of the performance of each program, with the accomplishment of objectives identified in the strategic plan constituting effective performance.

Increasing Reliability and Transparency of Annual and Periodic Assessment

Periodic assessment should be based on the same criteria as ongoing annual program assessment. Program attributes, activities, outcomes, and impacts that are sufficiently important to warrant annual or ongoing assessment are important enough to evaluate on a periodic basis. Review material prepared for the periodic review should be a compilation of the annual reports, book-ended by material that demonstrates the extent to which the annual activities combine to form a cohesive, ongoing program of activity organized to accomplish the objectives of an appropriately ambitious set of strategic plans and demonstrating effective progress towards accomplishment of the goals and objectives identified in those strategic plans.

Currently, the individual Sea Grant programs are ranked each year at the conclusion of the FE review. However, only one-quarter are actually rated in a given year (those that underwent a PAT review in the previous calendar year), thus the rankings change only in as much as the ratings for one-quarter of the programs changed in a given year. Thus the rankings reflect ratings that are as much as three years out of date for one-quarter of the programs, and three-quarters of the rating are at least one year out of date. The frequency of periodic assessment (once every four years) and the number of programs reviewed in a given year (one-quarter) is thus insufficient to support meaningful annual rankings of the programs as required by Congress. **The Director of the National Sea Grant College Program, in consultation with the National Sea Grant Review Panel and the directors of the individual Sea Grant programs, should modify the NSGO Final Evaluation review process so that every individual Sea Grant program is rated and ranked each year. The rating (and subsequent ranking) should be based on an assessment of each program's progress for the reporting year based on annual reports of activities, outcomes, and impacts in the context of the unique strategic plans approved for each program.** This is referred to as "Annual Assessment" in the Summary and Chapter 5, and is different from the current FE process.

Finally, as the functions of the annual and periodic assessments evolve, they will provide different and independent sources of information about the state of the program as a whole. This information should provide important insights about the status of the Sea Grant program overall to the Secretary of Commerce, the National Director, and potentially Congress. Thus, there is a need to synthesize and analyze the results of these assessments every four years, including a synthesis of the most recent periodic reviews of the individual programs and a systematic review of the NSGO. Developing such a "state of the program" report would seem to be an obvious role for the NSGRP. **The Director of the National Sea Grant College Program, acting under authority of the Secretary, should direct the National Sea Grant Review Panel to undertake the development of a systematic review of the "state of the Sea Grant program" once every four years. The review should rely extensively on information collected during the annual and periodic assessments, augmented with a site visit to the National Sea Grant Office, and should focus on how the program is functioning as a whole.** In addition to commenting on the how the program is performing in terms of the various criteria used during the assessments, the "state of the program" report could identify needed changes in program administration, conduct of the assessment process, or other areas as deemed valuable by the Secretary of Commerce or the National Director. The ability of the NSGRP to be seen as a credible source of such insight and advice to all parties may require that the NSGRP redefine its role in carrying out some components of the assessment. For example, greater consideration could be given to changing the NSGRP role to that of an observer, rather than actual evaluator, during the periodic assessments.

5

Major Findings and Recommendations

As pointed out in the previous chapters, the role of peer review and assessment within the National Sea Grant College Program (NSGCP or National Program) has evolved significantly since the program's inception, with many changes taking place since 1994. Collectively, there is evidence that these changes have led to a stronger program, although not all of the changes have been equally effective. In general, this report's analysis of efforts to "address the impact of the new procedures and evaluation process on Sea Grant as a whole" (see Box 1.2) suggests that changes in the evaluation process have been more successful in instituting competition as a mechanism for encouraging improvement in individual programs than in developing a national program that "provides an appropriately balanced response to local, regional, and national needs" (33 U.S.C. 1123). The following discussion summarizes the evolution of the evaluation process and makes recommendations for bringing greater balance to the evaluation process with regard both to appropriately directed competition and to development of a robust national program whose foundation is the network of local programs created and maintained by individual Sea Grant colleges and institutes and administered by the National Sea Grant Office (NSGO).

IMPACT OF CHANGES IN RESPONSE TO THE 1994 REPORT

Following the 1994 National Research Council (NRC) report *A Review of the NOAA National Sea Grant College Program*, the National Director instituted a number of changes in the way the program was evaluated.

Although it might be tempting to assume that simple quantitative measures such as publication counts would be useful, the value is marginal unless the collection and analysis of the information is carried out in a comprehensive manner. In order to carry out a direct assessment of the impacts of these changes on both the NSGCP and the individual Sea Grant programs, it would be necessary to conduct an independent assessment of the program, compare it to a similar pre-1994 assessment, and determine which differences are attributable to the changes related to the 1994 report. Even if it were possible to conduct such an exercise, doing so would be beyond the resources available during this study. There is, however, indirect evidence that the changes instituted after 1994 have strengthened the program.

First, one of the key recommendations of the 1994 report was to establish a process for strategic planning. Such a process was, in fact, established, and the individual Sea Grant programs have produced strategic plans. As discussed below, strategic planning within the NSGCP still needs to be improved, but on *prima facie* grounds, the adoption of a formal strategic planning process is an improvement over earlier practice.

Second, the current Sea Grant directors were asked whether the new evaluation process has led to improvements in their programs. The response was substantially in the affirmative. These responses cannot be taken as objective indicators of the effect of the new evaluation processes and must be interpreted with caution. On the other hand, the directors are by no means enthusiastic about all the details of the new process and were not reluctant to reveal this, so one might reasonably conclude that their responses to this particular question provide some useful information.

Third, other management practices employed by individual programs have demonstrably improved in direct response to periodic evaluation. These improvements include, but are not limited to, an enhanced relationship within the university administration, better internal reporting and accountability, a focus on documenting impacts and outcomes, and an awareness of long- and short-term goals. The prestige of the individual program is often increased by the visiting PAT members, leading to improved visibility and appreciation of the individual program within the administration of the home institution. In at least one case, the PAT report provided the necessary justification for creation of a full-time position to expand program efforts in education and outreach. The process of gathering materials necessary for a PAT visit also brought about increased effort for documenting impacts and outcomes. Many individual programs noted that the second visit was easier because they not only had an awareness of what materials were needed for the briefing books, but also had the opportunity to gather these materials during the years prior to the visit. In addition, the requirement (as mentioned above) to develop a strategic plan clearly aided the programs in terms of focusing the staff on goals, objectives, strategies, and outcomes. These responses from the individual program directors were tempered with concerns about the process, and some of the directors questioned if any of the acknowledged improvements were worth the expense and time that were invested in preparing for and hosting a PAT visit.

Finally, several members of the committee have first-hand, long-term experience with the Sea Grant program and it is their considered opinion that the changes instituted since 1994 have strengthened the program overall. As with the Sea Grant directors, the opinions of even knowledgeable individuals cannot be taken as objective indicators. But the unanimity of response to this issue—particularly in light of differences of opinions on other issues—suggests that real improvements have been made.

Effectiveness of Post-1998 Evaluation

As discussed in Chapter 4 and above, the most readily identified improvements in the NSGCP and the individual programs are directly attributable to administrative changes implemented in response to the 1994 NRC report and codified by The National Sea Grant College Program Reauthorization Act of 1998 (P.L. 105–160). The process subsequently established by the National Director and implemented by the NSGO to evaluate program performance and distribute merit funds as required by The National Sea Grant College Program Reauthorization Act of 2002 (P.L. 107–299) has also led to improvements in the overall program. However, several areas of concern remain.

Perhaps the foremost concern about the Sea Grant evaluation process is the reliance by the NSGO, working under the authority of the National Director, on periodic external assessments as the primary, if not only, means of evaluation and oversight. The periodic assessments are based largely on information collected during quadrennial visits by PATs overseen by the National Sea Grant Review Panel (NSGRP). Because the members of the PATs and the NSGRP are not federal employees, the preponderance of program evaluation is external. As the level of routine engagement of the NSGO with individual programs is rather low, reliance on external review reduces the federal component of the partnership that is central to the Sea Grant program. **The Director of the National Sea Grant College Program, under supervision of the Secretary of Commerce and in consultation with National Sea Grant Review Panel and the individual Sea Grant programs, should strengthen the ability of the National Sea Grant Office to carry out meaningful, ongoing internal assessment to complement periodic, external assessment currently tak-**

ing place. It is important to emphasize that this implies no criticism of the individuals who have participated as members of the PATs or the NSGRP, which operate in a highly professional and thorough manner. Program administration by the NSGO should make better use of annual reporting and regular interactions between the NSGO program officers and the directors of individual Sea Grant programs and administrators of their home institutions. These interactions should be centered on the development, approval, and implementation of strategic plans. The periodic, external reviews should continue because they provide an important opportunity to inject fresh perspectives and independent evaluation. Reverse site visits (see the LSAMP case study; Box 4.2) should be considered a potential mechanism for strengthening the connection between individual Sea Grant programs and the NSGO, allowing the perspectives offered by the individual programs to better shape the national and regional actions of the NSGO.

The reliance on the periodic assessments results in an unacceptable weighting of a single factor—the quadrennial PAT score—during the annual ranking of separate programs. The level of effort expended by all parties—the programs, the PAT members, and the NSGO—in evaluating a single program is so great that only 7 or 8 of the 31 programs have been assessed in any single year. Because the programs are ranked on an annual basis, the rankings are based on information that can be as much as 4 years out of date. As discussed in Chapter 3, the administrative rules established by the National Director (in partial response to P.L. 107–109) governing the distribution of merit funds, creates a situation in which closely ranked programs can receive substantially different awards (see Figure 3.4). The inherent subjectivity of the PAT evaluation, coupled with questions of reliability rooted in the minimal overlap of PAT membership, means that the PAT scores cannot be relied upon to discriminate the performance of different programs in a sufficiently meaningful way to justify relatively large differences in merit awards. While steps can and should be taken to further increase the reliability of the performance assessment process to support the rating and ranking of the individual programs, many of the changes proposed in this report may reduce the influence of the external periodic assessment process currently in use as a vehicle for identifying ways for the individual programs and the NSGO to work together to achieve the goal of providing "an appropriately balanced response to local, regional, and national needs" (33 U.S.C. 1123). The remainder of this chapter explores a number of changes that may be made to improve the overall value of program assessment within the Sea Grant program.

Strategic Planning

The importance of strategic planning in program development, implementation, and evaluation was emphasized in the 1994 NRC report. Specifically, the report recommended that "State Sea Grant Directors and the Director of the NSGO must cooperate to develop a single strategic plan articulating a shared vision and strategies which must be fully integrated into, and reflective of, NOAA's strategic plan" (NRC, 1994, p. 2). Although strategic planning at the national level (as carried out by the NSGO) meets this recommendation, the degree to which the national plan translates into action by individual programs is unclear. As recommended by the U.S. Commission on Ocean Policy, greater attention should be paid to regional scale issues. More effort is therefore needed to ensure that all of the individual programs develop strategic plans that are consistent with both national priorities, and local and regional priorities. To ensure that strategic planning reflects a shared vision, NSGO program officers should participate in the local strategic planning process, just as the directors of individual Sea Grant programs now participate in the development of the national plan. The strategic plan of each individual Sea Grant program should serve as the basis upon which that program is evaluated. **Steps should be taken by the Director of the National Sea Grant College Program, under supervision of the Secretary of Commerce and in consultation with National Sea Grant Review Panel and the individual Sea Grant programs, to strengthen strategic planning at both the national and individual program level. The strategic plans of the individual programs and the national program should represent a coordinated and collective effort to serve local, regional, and national needs.** As discussed in Chapter 4, actions by the National Director should include developing and implementing a process to assist individual programs in strategic planning, and creating a separate process for evaluating and approving appropriately ambitious strategic plans for the individual programs.

Performance Criteria

Performance criteria are a combination of quantitative and qualitative measures used to assess a selected program or activity, the program outcomes, and, in some instances, the system the program is intended to influence. In the case of assessing the effectiveness and impacts of individual Sea Grant programs, this involves setting benchmarks to describe the expected level of performance in a particular category (such as program organization and management) and indicators to help assess the performance of the individual program in that area. As discussed earlier, strategic planning is the critical basis for implementation, review, and evaluation of institutional programs. Yet at present, the strategic plans of each program are reviewed only as part of the periodic assessment of individual programs and concomitant with an assessment of the program's effectiveness in achieving the goals the plan describes. **The Director of the National Sea Grant College Program, under supervision of the Secretary of Commerce and in consultation with National Sea Grant Review Panel and the individual Sea Grant programs, should modify the benchmarks and indicators, as needed, to ensure that the performance of each program is measured against the objectives outlined in the separately approved, program specific strategic plan called for in the previous recommendation.**

In addition, the current Sea Grant evaluation criteria do not sufficiently recognize the importance of individual programs in building cooperative efforts to address regional and national scale problems. The existing benchmarks tend to encourage program development at the local scale. Furthermore, the heavy emphasis on individual program performance in determining merit and bonus allocations may have resulted in lower levels of cooperative behavior between programs, which now see themselves as pitted against one another. Encouraging programs to undertake cooperative efforts to address regional-scale problems needs to be incorporated into the evaluation process.

This call to modify the evaluation criteria to place greater weight on cooperative efforts is not intended as a recommendation to increase the complexity of the criteria. In the current review cycle (the assessment of all 30 programs over four years), 14 scored sub-criteria are considered in four major categories. As a consequence, considerable time and effort is devoted to assigning, and subsequently reviewing, a score in a criterion that may account for no more than 2 percent of the overall score. The current subdivision into 14 scored sub-criteria was not recommended by any of the major committees that have examined the process, nor is there evidence to suggest that 14 scored sub-criteria provide a more accurate assessment of program performance than a smaller number of less detailed criteria, as used in the first review cycle. **The Director of the National Sea Grant College Program, under supervision of the Secretary of Commerce and in consultation with National Sea Grant Review Panel and the individual Sea Grant programs, should substantially reduce the overall number of scored criteria by combining various existing criteria, while adding cooperative, network-building activities as an explicitly evaluated, highly valued criterion.** As discussed in Chapter 3, consideration should be given to reducing the number of scored criteria to be assessed in the next external, periodic review cycle. Rather than the existing 14 sub-criteria, ranging in weight from 2 percent to 25 percent, 4 to 6 broader criteria—weighted to reflect a balance among the production of meaningful results; outreach and education; planning; organization; management; and coordination among programs—would move assessment efforts toward more holistic judgments of program performance. Implementation of revised criteria should be postponed until the beginning of the next cycle of program review (the current review cycle will conclude in late 2006).

Program Assessment Team and Site Visit

Focusing the PAT visit on essential evaluation tasks would reduce the demand placed on PAT members and could allow members to participate in a larger number of reviews (thereby increasing reliability across evaluations) and reduce the cost of program assessment. Historically, the length and content of the PAT visits was largely determined through discussion between the director of the individual Sea Grant program under review and the chair of the PAT. Although the NSGCP has implemented changes to provide greater standardization, many individual Sea Grant programs have expressed concern that variability in program size (both in terms of geographic area covered and program budget and scope) requires significant flexibility in the length of the PAT visit and the amount of material provided to the PAT members.

No evidence was provided to substantiate concerns or claims that more complex (i.e., larger) programs required significantly longer PAT visits or greater volumes of supporting material. There is no reason to believe that greater standardization in the types and volume of information needed to characterize program performance would inappropriately handicap large programs. With regard to standardization of supporting material, it should be noted that the NSGO has made strides in the past year to reduce the amount and kinds of preparatory materials for PAT review. New language was added to the 2005 PAT Manual in the section called "PAT Preparation, Structure, and Cost Control" that provides suggestions for ways to minimize costs of the PAT visit, without reducing the PAT's effectiveness (NSGO, 2005a). This report supports these changes and suggests more of the same in the future. With regard to length of PAT visits, to some degree concerns in this area reflect the lack of clarity regarding what constitutes acceptable or exceptional performances in the various performance metrics used during the PAT process. **The Director of the National Sea Grant College Program, under supervision of the Secretary of Commerce and in consultation with National Sea Grant Review Panel and the individual Sea Grant programs, should shorten the duration and standardize the PAT site visits, based on the minimum time and material needed to cover essential, standardized ele-**

ments of the program assessment. The length of the PAT visit should be no more than the length of time needed to gather information and carry out a relatively uniform evaluation of all the programs using the modified metrical evaluation called for above. Based on the committee's experience, the essential information could be conveyed in two days, with a third day used for the PAT to complete its assessment and report out to the director and institutional representatives.

Providing Coordination and Facilitation Through Informed, Ongoing Oversight

Greater involvement and ongoing oversight by the NSGO is needed to ensure that the program as a whole continues to improve while addressing, local, regional and national needs. Informed oversight is also needed to lend credibility to annual program rankings and the allocation of merit and bonus funds. These two goals can be simultaneously served by a meaningful ongoing annual evaluation process that complements the periodic assessment carried out during the PAT review. Review material prepared for the periodic review should be a compilation of the annual reports of individual programs, supplemented by material that demonstrates the extent to which the annual activities combine to form a cohesive, ongoing program of activity organized to accomplish the objectives of an appropriately ambitious strategic plan and demonstrates effective progress towards accomplishment of the strategic plan's goals and objectives. **The Director of the National Sea Grant College Program, under supervision of the Secretary of Commerce, should rank the individual Sea Grant programs based on a program evaluation process that includes more robust, credible, and transparent annual assessments of each individual Sea Grant program.** Review of programs that have undergone periodic assessments in the preceding year should also include consideration of the PAT reports and the individual Sea Grant program directors' responses to the PAT reports. The additional effort required of individual Sea Grant programs to provide information on an annual basis can be offset to a degree by reducing the time required to prepare materials for the periodic review, if the majority of the information required by the latter is made up of materials submitted annually.

Fairness in Competition

Program ranking is often believed to be influenced by program size, age of the program, location, type of institutional administration linkages, term of the program officer, etc. With the exception of the term of the program officer with particular programs, statistical analysis failed to support these concerns. However, the current process produces a very narrow range of program scores, such that minute differences in assigned score may result in significant differences in the award of bonus funding. **The Director of the National Sea Grant College Program, under supervision of the Secretary of Commerce, should revise the calculation of bonus funding allocation relative to program rank to ensure that small differences in program rank do not result in large differences in bonus funding, while preserving or even enhancing the ability to competitively award bonus funds as required by the National Sea Grant College Program Act Amendments of 2002 (P.L. 107–299).** For example, as discussed in Chapter 3, the bonus pool could be distributed to the top half of the programs in proportion to the amount that each program's score exceeds that of the median-ranked program. Conversely, the amount of bonus funding could be increased uniformly by rank, so that each program eligible for bonus funding received an amount in proportion to its ranking.

Improving Program Cohesion

The NSGO does not currently play a sufficient role in ongoing program assistance, communication, and assessment, nor does it maintain close ongoing working relationships with the individual Sea Grant programs. This limits the ability of the NSGO, and by extension the National Director, to "provide an appropriately balanced response to local, regional, and national needs, which is reflective of integration with the relevant portions of strategic plans of the Department of Commerce and of the Administration"(33 U.S.C. 1123). There is a consensus among NSGO personnel and the directors of individual Sea Grant programs that there was a greater level of interaction between the NSGO and the individual Sea Grant programs prior to 1995. The expansion of external periodic review overseen by the NSGRP in partial response to successive amendments to 33 U.S.C. Chapter 22 has coincided with a reduced engagement by the NSGO in the ongoing activities of individual Sea Grant programs. As noted in Duce et al. (2002), closer and more frequent interaction with NSGO would help integrate individual Sea Grant programs and the National Program.

This reduced level of engagement by the NSGO staff appears to reflect several factors including:

- conflicting mandates to NSGO staff as part of broader efforts by NOAA to integrate functions across the organization,

- less emphasis on maintaining a high-level of interaction with individual programs by individual NSGO staff as individual programs assumed the responsibility for review of grant applications,
- a greater emphasis on external review of individual program performance, and
- turnover and attrition of the personnel in the NSGO.

In order for the NSGO to more effectively administer the program and coordinate and facilitate the efforts of the individual Sea Grant college and institutes, thus fulfilling the federal role within the Sea Grant partnership, the capabilities of the NSGO should be reevaluated and likely enhanced. **The Secretary of Commerce, in consultation with the National Sea Grant Review Panel, should take steps to ensure that sufficient human and fiscal resources are available to allow robust, ongoing, and meaningful interaction among the Director of the National Sea Grant College Program, the staff of the National Sea Grant Office, and the directors of individual Sea Grant programs, and the administrators of the institutional homes of the individual Sea Grant programs.**

This interaction will provide a solid foundation for the annual performance evaluation needed to annually rate and rank individual programs as required by law, and will help ensure that the various elements of the National Program are truly capable of providing "an appropriately balanced response to local, regional, and national needs, which is reflective of integration with the relevant portions of strategic plans of the Department of Commerce and of the Administration" (33 U.S.C. 1123). **The Director of the National Sea Grant College Program, under supervision of the Secretary of Commerce and in consultation with National Sea Grant Review Panel and the individual Sea Grant programs, should undertake an evaluation of how work force capabilities and other components of effective program administration could be modified within the National Sea Grant Office to enhance its ability to coordinate and facilitate the actions of the individual Sea Grant programs.** The implementation of changes in the NSGO that might be proposed from an evaluation will likely span many months or even years. In the interim, the performance of the NSGO could benefit from the type of external perspectives provided by bodies such as the PATs or the NSGRP. Site visits conducted by the PAT could provide a useful venue for such discussions and the resulting information could be channeled to NSGRP for further consideration.

Based on comments received during information gathering meetings hosted by the committee, written correspondence submitted in response to committee request, and various NSGO and NSGRP documents, it is apparent that an unacceptable number of individual Sea Grant program directors and their staff remain confused about key aspects of the periodic evaluation process, the annual evaluation process, and their impacts on program rankings and funding. Although responsibility for understanding this process rests with the individual Sea Grant program directors, the NSGO has a responsibility to make sure the process is reasonably straightforward and understandable. As discussed in Chapter 3, there should be greater attention and clarity regarding all aspects of program assessment. **The Director of the National Sea Grant College Program, under supervision of the Secretary of Commerce, should take steps to ensure that the program assessment process (both the new annual assessment called for in this report and the Program Assessment Team review) is well-described and understood by individual program directors, congressional staff, personnel at the Office of Management and Budget, university and state administrators, and the general public.**

If the recommendations put forward above are implemented, the functions of the annual and periodic assessments will evolve such that both will provide different and independent sources of information about the state of the program as a whole. This information should provide important insights about the state of Sea Grant program overall to the Secretary of Commerce, the National Director, and potentially Congress. Thus, there would seem to be a need to synthesize and analyze the results of these assessments every four years, including a synthesis of the most recent periodic reviews of the individual programs and a systematic review of the NSGO. Developing such a "state of the program" report would seem to be an obvious role for the NSGRP. **The Director of the National Sea Grant College Program, acting under authority of the Secretary, should direct the National Sea Grant Review Panel to undertake the development of a systematic review of the "state of the Sea Grant program" once every four years. The review should rely extensively on information collected during the annual and periodic reviews, augmented with a site visit to the National Sea Grant Office, and should focus on how the program is functioning as a whole.** In addition to commenting on the how the programs is performing in terms of the various criteria used during the assessments, the "state of the program" report could address needed changes in how the program is administered, how the assessment process is carried out, or other areas as deemed valuable by the Secretary or the National Director. The ability of the NSGRP to be seen as a credible source of such insight and advice to all parties may require evolution of the role of NSGRP in carrying out some components of the assessment. Greater consideration, for example, may need to be given to changing the NSGRP role to that of an observer, rather than the actual evaluator, during the periodic assessments.

References

Byrne, J.V., R.W. Corell, G.R. Heath, F. Kudrna, B. Baird, B.J. Copeland, F. Hutchinson, and M. Voiland. 2000. *A Mandate to Engage Coastal Users: A Review of the National Sea Grant College Extension Program and A Call for Greater National Commitment to Engagement*. The National Sea Grant Extension Review Panel, Oregon Sea Grant, Oregon State University, Corvallis, Oregon.

Copeland, B.J., B. Griswold, and C. Fetterolf. 1997. Evaluation of Sea Grant College Programs. Report to National Sea Grant College Program. Silver Spring, MD, 23 pp.

Council on Environmental Quality. 2005. *U.S. Ocean Action Plan*. Council on Environmental Quality, Washington, D.C. [Online] Available at: http://ocean.ceq.gov/actionplan.pdf [last accessed July 18, 2006].

CSREES (Cooperative State Research, Education and Extension Service). 1999. *Planning and Conducting Reviews: U.S. Department of Agriculture CSREES Guidelines for Institutions*. U.S. Department of Agriculture, Cooperative State Research, Education and Extension Service, Washington, D.C. [Online] Available at: http://www.csrees.usda.gov/business/review/csrstan.pdf [last accessed June 5, 2006].

Duce, R., M. Dubbs, L. Echols, F. Hutchinson, J. Schubel, J. Stephan and W. Stubblefield. 2002. *Building Sea Grant: The Role of the National Sea Grant Office*. Report of the National Sea Grant Review Panel. Silver Spring, Maryland, 89 pp.

Kudrna, F.L., P. Bell, E. Friday, M. Hernandez-Avila, N.E. Robinson, J.R. Stephan, and J.S. Weis. 2005. *Review and Recommendations: Sea Grant Program Evaluation Process*. Draft Report. Sea Grant Review Panel, Program Evaluation Committee, Silver Spring, Maryland. [Online] Available at: http://www.seagrant.noaa.gov/leadership/reviewpanel/evaluation_committee_final_draft_report_111705.pdf [last accessed July 14, 2006].

NRC (National Research Council). 1994. *A Review of the NOAA National Sea Grant College Program*. National Academy Press. Washington, D.C., 63 pp.

NRC. 2001. *The Measure of STAR: Review of the U.S. EPA Science to Achieve Results (STAR) Research Grants Program*. National Academies Press, Washington D.C.

NSF (National Science Foundation). 2003. *Louis Stokes Alliances for Minority Participation*. NSF 03-520. National Science Foundation, Directorate of Education and Human Resources, Arlington, Virginia. [Online] Available at: http://www.nsf.gov/pubs/2003/nsf03520/nsf03520.pdf [last accessed June 5, 2006].

NSGO (National Sea Grant Office). 1998. *Program Assessment Team (PAT) Visit Handbook*. National Sea Grant College Program, Silver Spring, MD, 15 pp.

NSGO. 1999a. *Program Assessment Team Manual*. National Sea Grant College Program, Silver Spring, Maryland, 28 pp.

NSGO. 1999b.National Sea Grant Program Memorandum on NSGO Final Evaluation and Merit Funding, April 22, 1999.

NSGO. 2000. *Program Assessment Team Manual*. National Sea Grant College Program, Silver Spring, Maryland, 28 pp.

NSGO. 2001. *Program Assessment Team Manual*. National Sea Grant College Program, Silver Spring, Maryland, 33 pp.

NSGO. 2003a. *Program Assessment Team Manual*. National Sea Grant College Program, Silver Spring, Maryland, 46 pp.

NSGO. 2003b. Policy for the Allocation of Funds, FY 2003 and Beyond. Memo of the National Office of Sea Grant. National Oceanic and Atmospheric Administration, Silver Spring, Maryland, 11 pp.

NSGO. 2004a. *Program Assessment Team Manual*. National Sea Grant College Program, Silver Spring, Maryland, 56 pp.

NSGO. 2004b. *NOAA Sea Grant Strategic Plan for FY 2003-2008 and Beyond*. National Sea Grant College Program, Silver Spring, Maryland, 36 pp.

NSGO. 2005a. *Program Assessment Team Manual*. National Sea Grant College Program, Silver Spring, Maryland, 59 pp.

NSGO. 2005b. Revised Policy Memorandum on NSGO Final Evaluation and Merit Funding (2005); April 8, 2005.

NSGO. 2005c. Summary of Noteworthy Highlights from the 2005 NSGO Final Evaluation and Merit Funding Meeting. February 7-11, 2005. Submitted by Frances Schuler to the Committee on the Evaluation of the Sea Grant Program Review Process.

Schuler, F. 2005. Presentation to the NRC Committee on Evaluation of the Sea Grant Program Review Process. National Research Council, Ocean Studies Board.

Spilhaus, A. 1964. Man in the Sea. *Science* 145(3636):993.

Toll, J., C. Fetterolf, J. Kramer, R. Malouf, N. Robinson, and J. Weis. 2001. *Review and Recommendations: Sea Grant Program Evaluation Process*. Report of the National Sea Grant Review Panel. National Sea Grant Review Panel, Silver Spring, Maryland, 84 pp.

USCOP (U.S. Commission on Ocean Policy). 2004. *An Ocean Blueprint for the 21st Century*. U.S. Commission on Ocean Policy, Washington, D.C.

Appendix A

Committee and Staff Biographies

COMMITTEE

Dr. James M. Coleman (*Chair*) received his Ph.D. in geology from Louisiana State University in 1966. He is the Boyd Professor for the Coastal Studies Institute of Louisiana State University and Agricultural and Mechanical College. Dr. Coleman is a former commissioner of the U.S. Commission on Ocean Policy, former chairman of the Marine Board, and former member of the Ocean Studies Board. He has served on many National Research Council committees. He is a member of the U.S. National Academy of Engineering and the Russian Academy of Natural Sciences. His research interests include coastal and marine processes and coastal management. He has received many awards in his nearly 40-year scientific career, including the Kapitsa Medal of Honor for his contributions to the field of petroleum sciences.

Mr. Robert J. Bailey earned his B.Sc. in earth science from Portland State University in 1968. He is manager of the Ocean and Coastal Services Division of the Oregon Department of Land Conservation and Development, Oregon's land-planning agency. Some of his duties include advising the Office of the Governor on matters of coastal and ocean policy, and program administration; representing the State of Oregon's coastal zone management interests during collaborations with federal agencies; and administering Oregon's Coastal Zone Management grant program (federally funded annually at $2 million). Mr. Bailey has also worked for several years as a land-use planner. Currently, he is the elected commissioner of the City of Oregon City. From 2001 to 2003, he was a he member of the City of Oregon City Planning Commission.

Dr. Billy J. (B.J.) Copeland received a Ph.D. from Oklahoma State University in 1963. He is Professor Emeritus in the Department of Zoology, North Carolina State University (NCSU). During his tenure at NCSU, he also functioned as Director of the North Carolina Sea Grant College from 1973 to 1996. His research interests include coastal ecology, water quality, habitat conservation, fisheries management, and natural resources development and conservation. Dr. Copeland served on the NRC National Science Foundation Graduate Panel on Biological Sciences. He has also served on numerous boards, committees, task forces, and commissions in the area of coastal and marine water quality and fisheries management. Aside from these activities, Dr. Copeland has authored roughly 150 articles, reports, book chapters and complete texts on matters related to coastal ecology. Currently, he serves on the North Carolina Marine Fisheries Commission.

Dr. Susan E. Cozzens received her Ph.D. in sociology from Columbia University. She is Professor of Public Policy at the Georgia Institute of Technology, and Director of its Technology Policy and Assessment Center. Dr. Cozzens has served on four NRC committees, including the Committee for Assessment of Centers of Excellence Programs at NIH and the Committee to Study the National Science Foundation Decisionmaking on Major Awards. Dr. Cozzens is the author of numerous articles in science policy and science and technology studies, and several books, including *Social Control and Multiple Discovery in Science: The Opiate Receptor Case* (SUNY Press, 1990), and *Theories of Science in Society* (coeditor with Thomas F. Gieryn; Indiana University Press, 1991). She is past editor of both *Science, Technology, & Human Values,* and *Society for Social Studies of Science* and current editor of *Research Evaluation*. Dr. Cozzens has served as a consultant to numerous organizations, including the Office of Science and Technology Policy, the National Science Foundation, and the National Institutes of Health. From 1995 through 1997, Dr. Cozzens was Director of the Office of Policy Support at the National Science Foundation.

Dr. Keith R. Criddle received his Ph.D. in agricultural economics from the University of California, Davis in 1989. He currently serves as the Ted Stevens Distinguished Professorship of Marine Policy at the University of Alaska Fairbanks. He was previously on the faculty in the Economics Departments of both at Utah State University in Logan, Utah, and at the University of Alaska Fairbanks. Dr. Criddle's research focuses on the intersection between the natural sciences and economics, especially the management of living resources. In particular, he has explored topics ranging from the economic impact of sport fishing in Cook Inlet, Alaska, to governance structures for fisheries management. Other research areas include sustainable fisheries management, fishery revenue maximizations, and evolution of the structure of the Chilean salmon aquaculture industry in response to requirements for traceability and assurance. Dr. Criddle served as the associate editor of Marine Resource Economics from 1993–2003 and as a member of the North Pacific Fisheries Management Council Scientific and Statistical Committee from 1993-present. He was a member of the NRC Committee on the Introduction of Nonnative Oysters in the Chesapeake Bay and on the NRC Committee to Review Individual Fishing Quotas.

Dr. Eliezer Geisler earned his Ph.D. from Northwestern University. He is Professor and Associate Dean at the Stuart Graduate School of Business of the Illinois Institute of Technology. Dr. Geisler is a leading scholar of the management of research, science and technology, and in knowledge management systems. He specializes in the management of healthcare and medical technology. His research, published in eight books and more than 100 scholarly articles, has contributed fundamental and innovative ideas that have significantly influenced the study of technology, R&D, science and knowledge in industry, universities and government, in the areas of technology alliances, entrepreneurship, commercialization of research, and technology transfer. In particular, Dr. Geisler is a leading scholar in the development of metrics for the evaluation of science and technology, and four of his books evaluate the impact of science.

Dr. Michael W. Howell earned his Ph.D. in marine science from the University of South Carolina and his M.S. in oceanic science from the University of Michigan. He is an associate professor at the University of South Florida; his research involves the use of deep-sea sediments to understand ocean and climate history through geological time. The paleoceanographic and paleoclimatic history of the Mediterranean Sea has been a major focus area of this work. Dr. Howell currently serves on the State of South Carolina Governor's Mathematics and Science Advisory Board, the American Geological Institute Minority Participation Program Advisory Committee, the Governing Board for the South Carolina Alliance for Minority Participation, the American Geophysical Union's Subcommittee on Diversity, and the Industrial Liaison Panel of the Integrated Ocean Drilling Program. Dr. Howell has served regularly on the National Science Foundation Directorate for Education and Human Resources, with a concentration in the Division of Elementary, Secondary, and Informal Education.

Mr. Richard C. Karney earned his B.Sc. in biological sciences from Rutgers University. Since 1976, he has been Shellfish Biologist and Director of the Martha's Vineyard Shellfish Group, Inc., a nonprofit consortium of the shellfish departments of six towns on Martha's Vineyard. Prior to 1976, Mr. Karney worked for the Virginia Institute of Marine Science. During his tenure at Martha's Vineyard Shellfish Group, he has carried out a successful community-based resource development program for the commercially important shellfish species on Martha's Vineyard. Management efforts have concentrated on the development of hatchery and field aquaculture methods for shellfish and the operation of the nation's first public solar shellfish hatchery. In the mid 1990's, with a $500,000 National Marine Fisheries Service grant, Mr. Karney conducted a shellfish aquaculture retraining program for fishermen displaced by fishing closures on Georges Bank. He is presently assisting the fishermen with marketing cultured oysters. Mr. Karney is also cochair of the National Shellfisheries Association Industry Subcommittee and cochair of the Southeast Massachusetts Aquaculture Center.

Dr. George I. Matsumoto received his Ph.D. in biological sciences from the University of California, Los Angeles, in 1990. Since 1996, he has been the Senior Educational and Research Specialist at the Monterey Bay Aquarium Research Institute (MBARI), Moss Landing, California. Dr. Matsumoto's research interests include open ocean and deep-sea communities; ecology and biogeography of open ocean and deep sea organisms; functional morphology, and natural history and behavior. In addition to research, his other responsibilities include managing several education and outreach efforts, including collaborations with MBARI's sister organization, the Monterey Bay Aquarium. Dr. Matsumoto served on the Digital Library for Earth System Education (DLESE) Steering Committee and the 2004 NSF Committee of Visitors for Geoscience Education and is currently serving on the Centers for Ocean Sciences Education Excellence (COSEE) National Advisory Board as well as Chair of the Ocean Research Interactive Observatory Networks (ORION) Education and Public Awareness Committee.

Dr. Joan Bray Rose received her Ph.D. in microbiology from the University of Arizona in 1985, and an M.S. in microbiology from the University of Wyoming in 1980. She joined the University of South Florida in April of 1989, first as associate and then as full professor, and recently accepted the Homer Nowlin Endowed Chair in Water Research at Michigan State University. Dr. Rose is an international expert in water pollution microbiology, waterborne disease, and public policy and health policy issues. Her research includes studies on waterborne diseases and microbial risk assessment. Her prior NRC service includes membership on the Water Science and Technology Board and the Board on Life Sciences and seven NRC committees. Dr. Rose is currently a member of the Science Advisory Board of the International Great Lakes Commission. She served as vice chair of the U.S. National Committee of the International Water Association from 2002 to 2004.

Dr. Andrew R. Solow earned his Ph.D. in geostatistics from Stanford University in 1986. He is Senior Scientist and Director of the Marine Policy Center at Woods Hole Oceanographic Institution. Dr. Solow's research experience involves environmental statistics, time-series analysis, spatial statistics, Bayesian methods, statistical biology, and ecology. He has authored some 150 scientific publications on topics including biological diversity, El Niño, and empirical analysis of volcanic eruptions. Dr. Solow is a former member of the NRC Commission on Geosciences, Environment, and Resources. He is currently serving on the Committee to Review the U.S. Ocean Research Priorities Plan and the Committee on Extending Observations and Research Results to Practical Applications: A Review of NASA's Approach.

Dr. Fred N. Spiess received his Ph.D. in physics from the University of California, Berkeley in 1951. He is currently a professor of oceanography, professor emeritus and research professor at Scripps Institution of Oceanography, University of California, San Diego. Dr. Spiess joined the Marine Physical Laboratory at Scripps in 1952 and began his research career in underwater acoustics and sonar systems. His interests include ocean engineering and related seagoing marine geophysics and graduate student education. From 1980 to 1988, Dr. Spiess was director of the University of California Institute of Marine Resources and was responsible for the administration of the California Sea Grant Program. Dr. Spiess has been a member of the National Academy of Engineering since 1985 as well as a member of the Ocean Studies Board. He has served on several NRC committees. Among other awards, Dr. Spiess holds the American Geophysical Union Ewing Medal and the Acoustical Society of America's Pioneers of Underwater Acoustics Medal.

NATIONAL RESEARCH COUNCIL STAFF

Dr. Dan Walker joined the National Research Council's Ocean Studies Board (OSB) as a program officer in July of 1995 and was named a National Academies' Scholar in 2005. Dr. Walker received his Ph.D. in Geology from the University of Tennessee in 1990. Prior to joining the OSB, Dan conducted research focused on the tectonic evolution of rifted continental margins, with an emphasis on natural resource (oil and gas) development and low-level radioactive waste disposal. Since joining the OSB, Dr. Walker has staffed nearly two dozen NRC studies, including *Earth Science and Applications from Space, Protecting and Restoring Coastal Louisiana, Evaluating the Sea Grant Review Process, Understanding Oil Spill Dispersants: Efficacy and Effects; Oil in the Sea III: Inputs, Fates, and Effects; Future Needs in Deep Submergence Science: Occupied and Unoccupied Vehicles in Basic Ocean Research; Environmental Information for Naval Warfare; Clean Coastal Waters: Understanding and Reducing the Effect of Nutrient Pollution; Science for Decisionmaking: Coastal and Marine Geology at the U.S. Geological Survey; Global Ocean Science: Toward an Integrated Approach*. Dr. Walker also directs the board's Engineering and Technology Subcommittee. He is also a guest investigator at the Marine Policy Center of the Woods Hole Oceanographic Institution and editor of the Marine Technology Society Journal.

Dr. Jennifer Merrill was a Senior Program Officer at the Ocean Studies Board from 2001 to 2005. She received her Ph.D. in Marine and Estuarine Environmental Science from the University of Maryland Center for Environmental Science, Horn Point Laboratory. Dr. Merrill served as a NOAA Knauss Marine Policy Fellow in the office of Senator Carl Levin, lectured at University of Maryland, and worked as a project manager at Maryland Sea Grant. At the OSB she directed studies that produced the reports *Marine Biotechnology in the Twenty-first Century: Problems, Promise, and Products* (2002), *Ocean Noise and Marine Mammals* (2003), *Exploration of the Seas: Voyage into the Unknown* (2003), and *Marine Mammal Populations and Ocean Noise: Determining When Noise Causes Biologically Significant Effects*. Dr. Merrill also assisted on the reports *Oil in the Sea III: Inputs, Fates, and Effects* (2003), and *Charting the Future of Methane Hydrates in the United States* (2004).

Ms. Amanda L. Babson was a National Academies Christine Mirzayan Science and Technology Policy Fellow at the Ocean Studies Board. Ms. Babson expects to be awarded her Ph.D. in Oceanography from the University of Washington in 2006. Ms. Babson received her B.A. in Physics from Carleton College in 1998.

Ms. Nancy Caputo is a research associate at the Ocean Studies Board, where she has worked since 2001. Ms. Caputo received an M.P.P. (Master of Public Policy) from the University of Southern California and a B.A. in political science/international relations from the University of California at Santa Barbara. Her interests include marine policy, science, and education. During her tenure with OSB, Ms. Caputo has assisted with the completion of seven reports: *A Review of the Florida Keys Carrying Capacity Study* (2002); *Emulsified Fuels—Risks and Response* (2002); *Decline of the Steller Sea Lion in Alaskan Waters—Untangling Food Webs and Fishing Nets* (2003); *Enabling Ocean Research in the 21st Century: Implementation of a Network of Ocean Observatories* (2003); *River Basins and Coastal Systems Planning Within the U.S. Army Corps of Engineers* (2004); *Charting the Future of Methane Hydrate Research in United States* (2004); *Dynamic Changes in Marine Ecosystems: Fishing, Food Webs, and Future Options* (2006). She is also the assistant editor of *Oceanography*, the professional magazine of The Oceanography Society.

Appendix B

List of Acronyms

ARS	Agricultural Research Service
CRIS	Current Research Information System
CSREES	Cooperative State Research, Education, and Extension Service
DOC	United States Department of Commerce
EPA	Environmental Protection Agency
FE	National Sea Grant Office Final Evaluation Review
FTE	Full Time Equivalent
FY	Fiscal Year
LSAMP	Louis Stokes Alliances for Minority Participation
MOS	Minimum Obligatory Set
NIH	National Institutes of Health
NOAA	National Oceanic and Atmospheric Administration
NRC	National Research Council
NRI	National Research Initiative (at USDA)
NSF	National Science Foundation
NSGCP	National Sea Grant College Program
NSGO	National Sea Grant Office (at NOAA)
NSGRP	National Sea Grant Review Panel

NSI	National Strategic Initiative
OAR	Ocean and Atmospheric Research (at NOAA)
OMB	Office of Management and Budget (Federal)
PA	Program Assessment
PAT	Program Assessment Team
PI	Principal Investigator
PO	Program Officer
RFP	Request for Proposal
SG	Sea Grant
SGA	Sea Grant Association
SMET	Natural sciences, mathematics, engineering and technology
STAR	Science to Achieve Results
TAT	Topical Advisory Team
USC	University of Southern California
USCOP	United States Commission on Ocean Policy
USDA	United States Department of Agriculture
WHOI	Woods Hole Oceanographic Institution

Appendix C

Key Sea Grant Legislation[1]

1 National Sea Grant College and Program Act of 1966 (P.L. 89–688)
2 National Sea Grant College Program Reauthorization Act of 1998 (P.L. 105–160)
3 National Sea Grant College Program Act Amendments of 2002 (P.L. 107–299)

---

[1] See Appendix H for U.S. Code, Title 33, Chapter 22.

PUBLIC LAW 89-688—OCTOBER 15, 1966 H.R. 16559

National Sea Grant College and Program Act of 1966

Public Law 89–688
October 15, 1966 (H.R. 16559)

An Act

To amend the Marine Resources and Engineering Development Act of 1966 to authorize the establishment and operation of sea grant colleges and programs by initiating and supporting programs of education and research in the various fields relating to the development of marine resources, and for other purposes.

National Sea Grant College and Program Act of 1966. Ant.p.203

Be it enacted by the Senate and House of Representatives of the United States of America in Congress assembled. That the Marine Resources and Engineering Development Act of 1966 is amended by adding at the end thereof the following new Title:

Title II—Sea Grant Colleges and Program
Short Title

Sec. 202. The Congress hereby finds and declares—

(a) that marine resources, including animal and vegetable life and mineral wealth, constitute a far-reaching and largely untapped asset of immense potential significance to the United States; and

(b) that it is in the national interest of the United States to develop the skilled manpower, including scientists, engineers, and technicians, and the facilities and equipment necessary for the exploitation of these resources; and

(c) that aquaculture, as with agriculture on land, and the gainful use of marine resources can substantially benefit the United States, and ultimately the people of the world, by providing greater economic opportunities, including expanded employment and commerce; the enjoyment and use of our marine resources; new sources of food; and new means for the development of marine resources; and

(d) that federal support toward the establishment, development, and operation of programs by sea grant colleges and federal support of other sea grant programs designed to achieve the gainful use of marine resources, offer the best means of promoting programs toward the goals set forth in clauses (a), (b), and (c), and should be undertaken by the federal government; and

(e) that in view of the importance of achieving the earliest possible institution of significant national activities related to the development of marine resources, it is the purpose of this title to provide for the establishment of a program of sea grant colleges and education, training, and research in the fields of marine science, engineering and related disciplines.

The provisions of this title shall be administered by the National Science Foundation Research programs, etc.

The foundation [the National Science Foundation] shall exercise its authority under this title by initiating and supporting programs at sea grant colleges and other suitable institutes, laboratories, and public or private agencies for the education of participants in the various fields relating to the development of marine resources with preference given to research aimed at practices, techniques, and design of equipment applicable to the development of marine resources; encouraging and developing programs consisting of instruction, practical demonstrations, publications with the object of imparting useful information to persons currently employed or interested in the various fields related to the development of marine resources.

The term "development of marine resources" means scientific endeavors relating to the marine environment, including but not limited to the fields oriented toward the development, conservation, or economic utilization of the physical, chemical, geological and biological resources of the marine environment, the fields of marine commerce and marine engineering, the fields relating to exploration or research in, the recover of natural resources from, and the transmission of energy in, the marine environment; the fields of oceanography and oceanology and the fields with respect to the study of the economic, legal, medical or sociological problems arising out of the management, use, development recovery and control of the natural resources of the marine environment. The term marine environment means the oceans, the Continental Shelf of the United States, the Great Lakes, the seabed and subsoil of the submarine areas adjacent to the coasts of the United States to the depth of two hundred meters or beyond that limit.

The term Sea Grant Program means any suitable public or private institution of higher education conducting any activities of education, research and advisory services oriented toward imparting information in fields related to the development of marine resources supported by the foundation.

Public Law 105–160
105th Congress

An Act

To reauthorize the Sea Grant Program.

Mar. 6, 1998
[S. 927]

Be it enacted by the Senate and House of Representatives of the United States of America in Congress assembled,

SECTION 1. SHORT TITLE.

National Sea Grant College Program Reauthorization Act of 1998. 33 USC 1121 note.

This Act may be cited as the "National Sea Grant College Program Reauthorization Act of 1998".

SEC. 2. AMENDMENT OF NATIONAL SEA GRANT COLLEGE PROGRAM ACT.

Except as otherwise expressly provided, whenever in this Act an amendment or repeal is expressed in terms of an amendment or repeal to, or repeal of, a section or other provision, the reference shall be considered to be made to a section or other provision of the National Sea Grant College Program Act (33 U.S.C. 1121 et seq.).

SEC. 3. FINDINGS.

(a) Section 202(a)(1) (33 U.S.C. 1121(a)(1)) is amended—

(1) by redesignating subparagraphs (D) and (E) as subparagraphs (E) and (F), respectively; and

(2) by inserting after subparagraph (C) the following:

"(D) encourage the development of forecast and analysis systems for coastal hazards;".

(b) Section 202(a)(6) (33 U.S.C. 1121(a)(6)) is amended by striking the second sentence and inserting the following: "The most cost-effective way to promote such activities is through continued and increased Federal support of the establishment, development, and operation of programs and projects by sea grant colleges, sea grant institutes, and other institutions.".

SEC. 4. DEFINITIONS.

(a) Section 203 (33 U.S.C. 1122) is amended—

(1) in paragraph (3)—

(A) by striking "their university or" and inserting "his or her"; and

(B) by striking "college, programs, or regional consortium" and inserting "college or sea grant institute";

(2) by striking paragraph (4) and inserting the following:

"(4) The term 'field related to ocean, coastal, and Great Lakes resources' means any discipline or field, including marine affairs, resource management, technology, education, or science, which is concerned with or likely to improve the understanding,

Resources of the House of Representatives and the Committee on Commerce, Science, and Transportation of the Senate.

(d) NOTICE OF REORGANIZATION.—The Secretary of Commerce shall provide notice to the Committees on Science, Resources, and Appropriations of the House of Representatives and the Committees on Commerce, Science, and Transportation and Appropriations of the Senate, not later than 45 days before any major reorganization of any program, project, or activity of the National Sea Grant College Program. 33 USC 1123 note.

SEC. 10. ADMINISTRATIVE LAW JUDGES. 15 USC 1541.

Notwithstanding section 559 of title 5, United States Code, with respect to any marine resource conservation law or regulation administered by the Secretary of Commerce acting through the National Oceanic and Atmospheric Administration, all adjudicatory functions which are required by chapter 5 of title 5 of such Code to be performed by an Administrative Law Judge may be performed by the United States Coast Guard on a reimbursable basis. Should the United States Coast Guard require the detail of an Administrative Law Judge to perform any of these functions, it may request such temporary or occasional assistance from the Office of Personnel Management pursuant to section 3344 of title 5, United States Code.

Approved March 6, 1998.

LEGISLATIVE HISTORY—S. 927:

SENATE REPORTS: No. 105–150 (Comm. on Commerce, Science, and Transportation).

CONGRESSIONAL RECORD:

Vol. 143 (1997): Nov. 13, considered and passed Senate.

Vol. 144 (1998): Feb. 11, considered and passed House, amended.

Feb. 12, Senate concurred in House amendment.

"(1) sea grant programs which comprise a national sea grant college program network, including international projects conducted within such programs;

"(2) administration of the national sea grant college program and this title by the national sea grant office, the Administration, and the panel;

"(3) the fellowship program under section 208; and

"(4) any national strategic investments in fields relating to ocean, coastal, and Great Lakes resources developed with the approval of the panel, the sea grant colleges, and the sea grant institutes.

"(c) RESPONSIBILITIES OF THE SECRETARY.—

"(1) The Secretary, in consultation with the panel, sea grant colleges, and sea grant institutes, shall develop a long-range strategic plan which establishes priorities for the national sea grant college program and which provides an appropriately balanced response to local, regional, and national needs.

"(2) Within 6 months of the date of enactment of the National Sea Grant College Program Reauthorization Act of 1998, the Secretary, in consultation with the panel, sea grant colleges, and sea grant institutes, shall establish guidelines related to the activities and responsibilities of sea grant colleges and sea grant institutes. Such guidelines shall include requirements for the conduct of merit review by the sea grant colleges and sea grant institutes of proposals for grants and contracts to be awarded under section 205, providing, at a minimum, for standardized documentation of such proposals and peer review of all research projects. Guidelines.

"(3) The Secretary shall by regulation prescribe the qualifications required for designation of sea grant colleges and sea grant institutes under section 207. Regulations.

"(4) To carry out the provisions of this title, the Secretary may—

"(A) appoint, assign the duties, transfer, and fix the compensation of such personnel as may be necessary, in accordance with civil service laws;

"(B) make appointments with respect to temporary and intermittent services to the extent authorized by section 3109 of title 5, United States Code;

"(C) publish or arrange for the publication of, and otherwise disseminate, in cooperation with other offices and programs in the Administration and without regard to section 501 of title 44, United States Code, any information of research, educational, training or other value in fields related to ocean, coastal, or Great Lakes resources;

"(D) enter into contracts, cooperative agreements, and other transactions without regard to section 5 of title 41, United States Code;

"(E) notwithstanding section 1342 of title 31, United States Code, accept donations and voluntary and uncompensated services;

"(F) accept funds from other Federal departments and agencies, including agencies within the Administration, to pay for and add to grants made and contracts entered into by the Secretary; and

"(G) promulgate such rules and regulations as may be necessary and appropriate.

"(d) DIRECTOR OF THE NATIONAL SEA GRANT COLLEGE PROGRAM.—

"(1) The Secretary shall appoint, as the Director of the National Sea Grant College Program, a qualified individual who has appropriate administrative experience and knowledge or expertise in fields related to ocean, coastal, and Great Lakes resources. The Director shall be appointed and compensated, without regard to the provisions of title 5, United States Code, governing appointments in the competitive service, at a rate payable under section 5376 of title 5, United States Code.

"(2) Subject to the supervision of the Secretary, the Director shall administer the national sea grant college program and oversee the operation of the national sea grant office. In addition to any other duty prescribed by law or assigned by the Secretary, the Director shall—

"(A) facilitate and coordinate the development of a long-range strategic plan under subsection (c)(1);

"(B) advise the Secretary with respect to the expertise and capabilities which are available within or through the national sea grant college program and encourage the use of such expertise and capabilities, on a cooperative or other basis, by other offices and activities within the Administration, and other Federal departments and agencies;

"(C) advise the Secretary on the designation of sea grant colleges and sea grant institutes, and, if appropriate, on the termination or suspension of any such designation; and

"(D) encourage the establishment and growth of sea grant programs, and cooperation and coordination with other Federal activities in fields related to ocean, coastal, and Great Lakes resources.

"(3) With respect to sea grant colleges and sea grant institutes, the Director shall—

"(A) evaluate the programs of sea grant colleges and sea grant institutes, using the priorities, guidelines, and qualifications established by the Secretary;

"(B) subject to the availability of appropriations, allocate funding among sea grant colleges and sea grant institutes so as to—

"(i) promote healthy competition among sea grant colleges and institutes;

"(ii) encourage successful implementation of sea grant programs; and

"(iii) to the maximum extent consistent with other provisions of this Act, provide a stable base of funding for sea grant colleges and institutes; and

"(C) ensure compliance with the guidelines for merit review under subsection (c)(2).".

SEC. 6. REPEAL OF SEA GRANT INTERNATIONAL PROGRAM.

Section 3 of the Sea Grant Program Improvement Act of 1976 (33 U.S.C. 1124a) is repealed.

SEC. 7. SEA GRANT COLLEGES AND SEA GRANT INSTITUTES.

Section 207 (33 U.S.C. 1126) is amended to read as follows:

"SEC. 207. SEA GRANT COLLEGES AND SEA GRANT INSTITUTES.

"(a) DESIGNATION.—

"(1) A sea grant college or sea grant institute shall meet the following qualifications—

"(A) have an existing broad base of competence in fields related to ocean, coastal, and Great Lakes resources;

"(B) make a long-term commitment to the objective in section 202(b), as determined by the Secretary;

"(C) cooperate with other sea grant colleges and institutes and other persons to solve problems or meet needs relating to ocean, coastal, and Great Lakes resources;

"(D) have received financial assistance under section 205 of this title (33 U.S.C. 1124);

"(E) be recognized for excellence in fields related to ocean, coastal, and Great Lakes resources (including marine resources management and science), as determined by the Secretary; and

"(F) meet such other qualifications as the Secretary, in consultation with the panel, considers necessary or appropriate.

"(2) The Secretary may designate an institution, or an association or alliance of two or more such institutions, as a sea grant college if the institution, association, or alliance—

"(A) meets the qualifications in paragraph (1); and

"(B) maintains a program of research, advisory services, training, and education in fields related to ocean, coastal, and Great Lakes resources.

"(3) The Secretary may designate an institution, or an association or alliance of two or more such institutions, as a sea grant institute if the institution, association, or alliance—

"(A) meets the qualifications in paragraph (1); and

"(B) maintains a program which includes, at a minimum, research and advisory services.

"(b) EXISTING DESIGNEES.—Any institution, or association or alliance of two or more such institutions, designated as a sea grant college or awarded institutional program status by the Director prior to the date of enactment of the National Sea Grant College Program Reauthorization Act of 1998, shall not have to reapply for designation as a sea grant college or sea grant institute, respectively, after the date of enactment of the National Sea Grant College Program Reauthorization Act of 1998, if the Director determines that the institution, or association or alliance of institutions, meets the qualifications in subsection (a).

"(c) SUSPENSION OR TERMINATION OF DESIGNATION.—The Secretary may, for cause and after an opportunity for hearing, suspend or terminate any designation under subsection (a).

"(d) DUTIES.—Subject to any regulations prescribed or guidelines established by the Secretary, it shall be the responsibility of each sea grant college and sea grant institute—

"(1) to develop and implement, in consultation with the Secretary and the panel, a program that is consistent with the guidelines and priorities established under section 204(c); and

"(2) to conduct a merit review of all proposals for grants and contracts to be awarded under section 205.".

SEC. 8. SEA GRANT REVIEW PANEL.

(a) Section 209(a) (33 U.S.C. 1128(a)) is amended by striking the second sentence.

(b) Section 209(b) (33 U.S.C. 1128(b)) is amended—

(1) by striking "The Panel" and inserting "(b) DUTIES.—The panel";

(2) by striking "and section 3 of the Sea Grant College Program Improvement Act of 1976" in paragraph (1); and

(3) by striking "regional consortia" in paragraph (3) and inserting "institutes".

(c) Section 209(c) (33 U.S.C. 1128(c)) is amended—

(1) in paragraph (1) by striking "college, sea grant regional consortium, or sea grant program" and inserting "college or sea grant institute"; and

(2) by striking paragraph (5)(A) and inserting the following:

"(A) receive compensation at a rate established by the Secretary, not to exceed the maximum daily rate payable under section 5376 of title 5, United States Code, when actually engaged in the performance of duties for such panel; and".

SEC. 9. AUTHORIZATION OF APPROPRIATIONS.

(a) GRANTS, CONTRACTS, AND FELLOWSHIPS.—Section 212(a) (33 U.S.C. 1131(a)) is amended to read as follows:

"(a) AUTHORIZATION.—

"(1) IN GENERAL.—There is authorized to be appropriated to carry out this Act—

"(A) $56,000,000 for fiscal year 1999;

"(B) $57,000,000 for fiscal year 2000;

"(C) $58,000,000 for fiscal year 2001;

"(D) $59,000,000 for fiscal year 2002; and

"(E) $60,000,000 for fiscal year 2003.

"(2) ZEBRA MUSSEL AND OYSTER RESEARCH.—In addition to the amount authorized for each fiscal year under paragraph (1)—

"(A) up to $2,800,000 may be made available as provided in section 1301(b)(4)(A) of the Nonindigenous Aquatic Nuisance Prevention and Control Act of 1990 (16 U.S.C. 4741(b)(4)(A)) for competitive grants for university research on the zebra mussel;

"(B) up to $3,000,000 may be made available for competitive grants for university research on oyster diseases and oyster-related human health risks; and

"(C) up to $3,000,000 may be made available for competitive grants for university research on Pfiesteria piscicida and other harmful algal blooms.".

(b) LIMITATION ON CERTAIN FUNDING.—Section 212(b)(1) (33 U.S.C. 1131(b)(1)) is amended to read as follows:

"(b) PROGRAM ELEMENTS.—

"(1) LIMITATION.—No more than 5 percent of the lesser of—

"(A) the amount authorized to be appropriated; or

"(B) the amount appropriated,

for each fiscal year under subsection (a) may be used to fund the program element contained in section 204(b)(2).".

33 USC 1131 note.

(c) NOTICE OF REPROGRAMMING.—If any funds authorized by this section are subject to a reprogramming action that requires notice to be provided to the Appropriations Committees of the House of Representatives and the Senate, notice of such action shall concurrently be provided to the Committees on Science and

Resources of the House of Representatives and the Committee on Commerce, Science, and Transportation of the Senate.

(d) NOTICE OF REORGANIZATION.—The Secretary of Commerce shall provide notice to the Committees on Science, Resources, and Appropriations of the House of Representatives and the Committees on Commerce, Science, and Transportation and Appropriations of the Senate, not later than 45 days before any major reorganization of any program, project, or activity of the National Sea Grant College Program. 33 USC 1123 note.

SEC. 10. ADMINISTRATIVE LAW JUDGES. 15 USC 1541.

Notwithstanding section 559 of title 5, United States Code, with respect to any marine resource conservation law or regulation administered by the Secretary of Commerce acting through the National Oceanic and Atmospheric Administration, all adjudicatory functions which are required by chapter 5 of title 5 of such Code to be performed by an Administrative Law Judge may be performed by the United States Coast Guard on a reimbursable basis. Should the United States Coast Guard require the detail of an Administrative Law Judge to perform any of these functions, it may request such temporary or occasional assistance from the Office of Personnel Management pursuant to section 3344 of title 5, United States Code.

Approved March 6, 1998.

LEGISLATIVE HISTORY—S. 927:

SENATE REPORTS: No. 105–150 (Comm. on Commerce, Science, and Transportation).

CONGRESSIONAL RECORD:

Vol. 143 (1997): Nov. 13, considered and passed Senate.

Vol. 144 (1998): Feb. 11, considered and passed House, amended.

Feb. 12, Senate concurred in House amendment.

Public Law 107–299
107th Congress

An Act

To reauthorize the National Sea Grant College Program Act, and for other purposes.

Nov. 26, 2002
[H.R. 3389]

Be it enacted by the Senate and House of Representatives of the United States of America in Congress assembled,

National Sea Grant College Program Act Amendments of 2002.
33 USC 1121 note.

SECTION 1. SHORT TITLE.

This Act may be cited as the "National Sea Grant College Program Act Amendments of 2002".

SEC. 2. AMENDMENTS TO FINDINGS.

Section 202(a)(6) of the National Sea Grant College Program Act (33 U.S.C. 1121(a)(6)) is amended by striking the period at the end and inserting ", including strong collaborations between Administration scientists and scientists at academic institutions.".

SEC. 3. REQUIREMENTS APPLICABLE TO NATIONAL SEA GRANT COLLEGE PROGRAM.

(a) Quadrennial Strategic Plan.—Section 204(c)(1) of the National Sea Grant College Program Act (33 U.S.C. 1123(c)(1)) is amended to read as follows:

"(1) The Secretary, in consultation with the panel, sea grant colleges, and sea grant institutes, shall develop at least every 4 years a strategic plan that establishes priorities for the national sea grant college program, provides an appropriately balanced response to local, regional, and national needs, and is reflective of integration with the relevant portions of the strategic plans of the Department of Commerce and of the Administration.".

(b) Program Evaluation and Rating.—

(1) Evaluation and rating requirement.—Section 204(d)(3)(A) of the National Sea Grant College Program Act (33 U.S.C. 1123(d)(3)(A)) is amended to read as follows:

"(A)(i) evaluate the performance of the programs of sea grant colleges and sea grant institutes, using the priorities, guidelines, and qualifications established by the Secretary under subsection (c), and determine which of the programs are the best managed and carry out the highest quality research, education, extension, and training activities; and

"(ii) rate the programs according to their relative performance (as determined under clause (i)) into no less than 5 categories, with each of the 2 best-performing categories containing no more than 25 percent of the programs;".

Contracts. 33 USC 1123 note.

(2) REVIEW OF EVALUATION AND RATING PROCESS.—(A) After 3 years after the date of the enactment of this Act, the Secretary of Commerce, acting through the Under Secretary of Commerce for Oceans and Atmosphere, shall contract with the National Academy of Sciences—

(i) to review the effectiveness of the evaluation and rating system under the amendment made by paragraph (1) in determining the relative performance of programs of sea grant colleges and sea grant institutes;

(ii) to evaluate whether the sea grant programs have improved as a result of the evaluation process; and

(iii) to make appropriate recommendations to improve the overall effectiveness of the evaluation process.

Reports. Deadline.

(B) The National Academy of Sciences shall submit a report to the Congress on the findings and recommendations of the panel under subparagraph (A) by not later than 4 years after the date of the enactment of this Act.

(c) ALLOCATION OF FUNDING.—Section 204(d)(3)(B) of the National Sea Grant College Program Act (33 U.S.C. 1123(d)(3)(B)) is amended by striking "and" after the semicolon at the end of clause (ii) and by adding at the end the following:

"(iv) encourage and promote coordination and cooperation between the research, education, and outreach programs of the Administration and those of academic institutions; and".

SEC. 4. COST SHARE.

Section 205(a) of the National Sea Grant College Program Act (33 U.S.C. 1124(a)) is amended by striking "section 204(d)(6)" and inserting "section 204(c)(4)(F)".

SEC. 5. FELLOWSHIPS.

(a) ENSURING EQUAL ACCESS.—Section 208(a) of the National Sea Grant College Program Act (33 U.S.C. 1127(a)) is amended by adding at the end the following: "The Secretary shall strive to ensure equal access for minority and economically disadvantaged students to the program carried out under this subsection. Not later than 1 year after the date of the enactment of the National Sea Grant College Program Act Amendments of 2002, and every 2 years thereafter, the Secretary shall submit a report to the Congress describing the efforts by the Secretary to ensure equal access for minority and economically disadvantaged students to the program carried out under this subsection, and the results of such efforts.".

Deadline. Reports.

(b) POSTDOCTORAL FELLOWS.—Section 208(c) of the National Sea Grant College Program Act (33 U.S.C. 1127(c)) is repealed.

SEC. 6. TERMS OF MEMBERSHIP FOR SEA GRANT REVIEW PANEL.

Section 209(c)(2) of the National Sea Grant College Program Act (33 U.S.C. 1128(c)(2)) is amended by striking the first sentence and inserting the following: "The term of office of a voting member of the panel shall be 3 years for a member appointed before the date of enactment of the National Sea Grant College Program Act Amendments of 2002, and 4 years for a member appointed or reappointed after the date of enactment of the National Sea Grant College Program Act Amendments of 2002. The Director may extend the term of office of a voting member of the panel appointed before the date of enactment of the National Sea Grant College Program Act Amendments of 2002 by up to 1 year.”.

SEC. 7. AUTHORIZATION OF APPROPRIATIONS.

Subsections (a), (b), and (c) of section 212 of the National Sea Grant College Program Act (33 U.S.C. 1131) are amended to read as follows:

“(a) AUTHORIZATION.—

“(1) IN GENERAL.—There are authorized to be appropriated to the Secretary to carry out this title—

“(A) $60,000,000 for fiscal year 2003;

“(B) $75,000,000 for fiscal year 2004;

“(C) $77,500,000 for fiscal year 2005;

“(D) $80,000,000 for fiscal year 2006;

“(E) $82,500,000 for fiscal year 2007; and

“(F) $85,000,000 for fiscal year 2008.

“(2) PRIORITY ACTIVITIES.—In addition to the amounts authorized under paragraph (1), there are authorized to be appropriated for each of fiscal years 2003 through 2008—

“(A) $5,000,000 for competitive grants for university research on the biology and control of zebra mussels and other important aquatic nonnative species;

“(B) $5,000,000 for competitive grants for university research on oyster diseases, oyster restoration, and oyster-related human health risks;

“(C) $5,000,000 for competitive grants for university research on the biology, prevention, and forecasting of harmful algal blooms, including Pfiesteria piscicida; and

“(D) $3,000,000 for competitive grants for fishery extension activities conducted by sea grant colleges or sea grant institutes to enhance, and not supplant, existing core program funding.

“(b) LIMITATIONS.—

“(1) ADMINISTRATION.—There may not be used for administration of programs under this title in a fiscal year more than 5 percent of the lesser of—

“(A) the amount authorized to be appropriated under this title for the fiscal year; or

“(B) the amount appropriated under this title for the fiscal year.

“(2) USE FOR OTHER OFFICES OR PROGRAMS.—Sums appropriated under the authority of subsection (a)(2) shall not be available for administration of this title by the National Sea Grant Office, for any other Administration or department program, or for any other administrative expenses.

“(c) DISTRIBUTION OF FUNDS.—In any fiscal year in which the appropriations made under subsection (a)(1) exceed the amounts appropriated for fiscal year 2003 for the purposes described in such subsection, the Secretary shall distribute any excess amounts (except amounts used for the administration of the sea grant program) to any combination of the following:

“(1) sea grant programs, according to their rating under section 204(d)(3)(A);

“(2) national strategic investments authorized under section 204(b)(4);

"(3) a college, university, institution, association, or alliance for activities that are necessary for it to be designated as a sea grant college or sea grant institute; and

"(4) a sea grant college or sea grant institute designated after the date of enactment of the National Sea Grant College Program Act Amendments of 2002 but not yet evaluated under section 204(d)(3)(A).".

SEC. 8. ANNUAL REPORT ON PROGRESS IN BECOMING DESIGNATED AS SEA GRANT COLLEGES AND SEA GRANT INSTITUTES.

Section 207 of the National Sea Grant College Program Act
33 USC 1126. (16 U.S.C. 1126) is amended by adding at the end the following:

"(e) ANNUAL REPORT ON PROGRESS.—

"(1) REPORT REQUIREMENT.—The Secretary shall report annually to the Committee on Resources and the Committee on Science of the House of Representatives, and to the Committee on Commerce, Science, and Transportation of the Senate, on efforts and progress made by colleges, universities, institutions, associations, and alliances to become designated under this section as sea grant colleges or sea grant institutes, including efforts and progress made by sea grant institutes in being designated as sea grant colleges.

"(2) TERRITORIES AND FREELY ASSOCIATED STATES.—The report shall include description of—

"(A) efforts made by colleges, universities, associations, institutions, and alliances in United States territories and freely associated States to develop the expertise necessary to be designated as a sea grant institute or sea grant college;

"(B) the administrative, technical, and financial assistance provided by the Secretary to those entities seeking to be designated; and

"(C) the additional actions or activities necessary for those entities to meet the qualifications for such designation under subsection (a)(1).".

33 USC 857–20. **SEC. 9. COORDINATION.**

Deadline. Reports.

Not later than February 15 of each year, the Under Secretary of Commerce for Oceans and Atmosphere and the Director of the National Science Foundation shall jointly submit to the Committees on Resources and Science of the House of Representatives and the Committee on Commerce, Science, and Transportation of the Senate a report on how the oceans and coastal research activities of the National Oceanic and Atmospheric Administration, including the Coastal Ocean Program and the National Sea Grant College Program, and of the National Science Foundation will be coordinated during the fiscal year following the fiscal year in which the report is submitted. The report shall describe in detail any overlapping ocean and coastal research interests between the agencies and specify how such research interests will be pursued by the programs in a complementary manner.

Approved November 26, 2002.

LEGISLATIVE HISTORY—H.R. 3389 (S. 2428):

HOUSE REPORTS: No. 107–369, Pt. 1 (Comm. on Resources) and Pt. 2 (Comm. on Science).
SENATE REPORTS: No. 107–187 accompanying S. 2428 (Comm. on Commerce, Science, and Transportation).
CONGRESSIONAL RECORD, Vol. 148 (2002):
June 19, considered and passed House.
Oct. 10, considered and passed Senate, amended.
Nov. 12, House concurred in Senate amendment.

○

Appendix D

National Sea Grant Program Memorandum on NSGO Final Evaluation and Merit Funding, April 22, 1999

[Stamped Date on Original: April 22, 1999]

MEMORANDUM FOR: Sea Grant Directors

FROM: Ronald C. Baird
Director

SUBJECT: Policy Memorandum on NSGO Final Evaluation and Merit Funding

Introduction

The National Sea Grant Office (NSGO) guidelines set forth in the policy document of April 20, 1998, "Implementation of Program Evaluation Procedures in the National Sea Grant College Program," require that the NSGO submit to each institution responsible for administering a Sea Grant Program, a Final Evaluation and Recommendation Report that summarizes the findings of a performance review for that Sea Grant Program over a four-year program cycle. The primary objectives of the Report are to provide local management with an assessment of performance and specific recommendations directed toward performance improvement and maintenance of existing program strengths. This memorandum establishes the procedures of the NSGO in the conduct of the review and preparation of the report.

In addition to the Report, and in accordance with the policy document of April 20, 1998 the NSGO is instructed to provide a performance rating to each Sea Grant program being evaluated. The rating is the basis for allocating merit-based funding from a pool of funds set aside in the Sea Grant budget for that purpose. This memorandum also establishes procedures and mechanics for rating determination and the allocation of merit funding.

NSGO Review Process

General: The NSGO will conduct the final review of a Sea Grant program within one year following the Program Assessment Team (PAT) site visit. All programs are evaluated to the extent possible in a similar manner and against common performance benchmarks. It must be recognized, however, that improvements in methodology in evaluating performance can be expected to be incorporated into the process over time.

The final evaluation review will be conducted by the NSGO technical staff and one or more members of the Executive Committee of the National Sea Grant Review Panel. The review is expected to be completed over a one week period during February/March of each year. All programs having had PAT visits in the previous calendar year will be considered as a group so as to maintain a consistent four-year review cycle. For example, the NSGO review session for 1999 occurred from February 11-19 and evaluations of the eight programs having PAT reviews in calendar year 1998 were included.

After each program has been evaluated, the NSGO will prepare the report that will contain the principal findings of the review and recommendations for enhancement of program performance. The report and a summary cover letter from the director of the NSGO will be sent to the director of each program being evaluated. The covering letter and report will also be copied to the appropriate administrators of the institution with administrative responsibility for that Sea Grant program. The report will normally be submitted prior to the end of April in each calendar year.

The following sections explain in greater detail the materials used and the structure of the review process.

Reports and Information Considered: In preparation for the final review, the NSGO Program Officer prepares for distribution materials that will be considered in the review. The compendium of documents and reports made available to NSGO staff for study prior to the review are listed below. Some of these materials are not generally available to the PAT and represent additional information for the NSGO to use in the evaluation process.

- The PAT report along with the institution's response
- The program's strategic plan
- Pertinent descriptive material from implementation plans/omnibus proposals etc.
- Annual progress reports
- Information on major accomplishments
- Trip and peer review panel reports by the Program Officer
- Topical Assessment Team reports (if any)
- Sea Grant funding information
- Other material deemed to be relevant by the Program Officer

Review Structure: The director of the NSGO facilitates the review sessions. Each session is limited to a discussion of one program only. Thus each program is reviewed on its own merits and not in competition with other programs. The NSGO Program Officer provides a brief overview of the program's performance over the last four years. Information is presented about the Sea Grant program's management structure and institutional setting, its outside advisory and strategic planning process, the programmatic areas of emphasis, major program accomplishments and outstanding issues or concerns.

The NSGO final review attempts to amplify and add to the information base produced in connection with the PAT site visit. The evaluation report prepared by the PAT however provides a significant input into the NSGO's final evaluation. The program's response to the PAT report and any recent information received about corrective actions are also considered. Evaluation-related materials on file in the NSGO, including progress reports or the response to recommendations from previous reviews, are weighed. Participation in national initiatives and responsiveness to network-wide activities are also considered. These bear on successful performance, and usually the NSGO has a better perspective here than the PAT.

Each of the four major evaluation criteria is discussed in sequence including the PAT findings. All PAT recommendations are reviewed. Those deemed most critical are highlighted for inclusion in the report. Where appropriate, PAT recommendations may be modified and additional recommendations developed based on the NSGO review. During the review process "Best Management Practices" are identified. These will be summarized for later promulgation to the Sea Grant network.

The discussions and findings from the review form the basis for the report that is prepared under the direction of the director NSGO.

Merit Fund Ratings and Allocations

General: In accordance with NSGO guidelines, there has been established in the Sea Grant budget a pool of funds to be allocated to individual Sea Grant programs on the basis of overall performance. It is the responsibility of the NSGO to develop a system to rate each Sea Grant program for the purpose of allocating funds from this merit pool. The NSGO is also charged with providing a rationale and mechanism for the disbursement of such funds in accordance with the rating system while considering both variations in the amount of funding available to the pool as well as changes in the distribution of program ratings that can occur each year. The rationale and protocols for both the rating system and allocation of merit funding are presented in the next section.

Merit Categories and Decision Protocols: The evaluation of program performance involves the use of judgment in weighing the qualitative and quantitative evidence available to the NSGO. It is also recognized that better metrics for the measurement of performance will be developed over time and incorporation of these must be provided for in the rating scheme. For the purpose of merit funding, the NSGO believes it is neither possible nor advisable to create a rating system that attempts to determine fine scale differences in performance among structurally dissimilar programs. Consequently, the system developed consists of only four rating categories of satisfactory or better performance and one default category for unsatisfactory performance, assignment to which involves substantial corrective action and could lead to reductions in future funding allocations.

After the evaluation process, programs are assigned to one of the four rating categories. Ratings are based on grading of the same four criteria as the PAT evaluations and carry the same weighting percentages. The PAT assessments are significant determinants in arriving at an overall final merit rating even though the NSGO categories are not directly comparable to those of the PAT. For instance, programs that did not carry a PAT grade of "Excellent" for the criteria "Producing Significant Results" were not rated in the highest performance categories (1 and 2) by the NSGO in the 1999 evaluations.

Final program ratings then reflect a consensus from both the NSGO and outside PAT based on a collective assessment of how well a program has performed in relation to the evaluation criteria over the last four years. Categories 1 and 2 are reserved for programs that achieve the highest levels of performance. Category 3 denotes programs meeting performance benchmarks, while programs assigned to Category 4 have significant deficiencies.

Merit Pool Categories: Programs assigned to categories 1, 2, and 3 qualify for merit pool allocations over the next four years. Programs assigned to Category 4 will not receive a merit pool allocation during the four-year period. The merit pool allocation will consist of two parts, a minimum allocation that is fixed and available annually for each of four years, and a residual share component that is variable and may change each year depending upon the performance ratings of all programs that have been reviewed. Added together, these two components—the minimum allocation and the residual share—determine each program's merit funding allocation for a given year. The merit pool allocations are set up in this manner so that all programs will have the same merit funding opportunities, regardless of the year they are reviewed.

Minimum Allocation: The minimum allocation is a fixed percentage of the merit pool that a program can expect to receive over the course of the next four years. Assuming level funding in the pool this will be a constant amount annually. The fixed minimum component for a program in Category 1 is calculated by dividing the merit pool by the number of programs that have been reviewed (8 in 1999, 29 after four years). A program in Category 2 and Category 3 would receive a minimum allocation of 70 percent and 40 percent respectively of that received by a program in Category 1. For example, if the total merit pool for 29 programs is $2,900,000, the minimum component would be $100,000, $70,000, and $40,000 for a program in category 1, 2, and 3 respectively. A program in Category 4 would not receive an allocation.

Once the amount in the merit pool is determined, the fixed minimum component remains unchanged until the total merit pool amount changes. Such a change could result from changes in the Sea Grant program's appropriation or a change in budget strategy. Our goal is to eventually have up to 10 percent of the core funding allocated through the merit pool. In 1999 the funds used for regional and multiprogram activities, $2,900,000 of which are being phased into the merit pool, amount to about 7 percent of core funding. Note also that budget considerations might require reductions in the funding of the merit pool sometime in the future.

Residual Share: The residual share is a variable component that depends upon how much of the merit pool remains unallocated after covering the fixed minimum allocations. It also depends on the distribution of ratings across all programs. The amount that remains after meeting the minimum allocations, the _residual_ amount, is distributed to programs in Category 1 and Category 2 only. Category 1 gets twice as much as Category 2. A new residual share is calculated every year. The Director of the NSGO could cap an award if the residual share exceeds 10 percent of the merit pool, although in practice this is unlikely to occur.

Program Merit Pool Allocations: The fixed minimum allocation and the residual share are added together to determine the merit pool allocations for each program. In 1999, only eight programs had been reviewed and these will receive the merit funding allocations in FY2000. By the end of four years, all programs will have been evaluated and the review process cycle will have reached steady state. The attached charts show the details of the calculations and several hypothetical examples of how merit pool allocations are likely to change from year-to-year due to the variable residual share component.

Restrictions on Allocated Funds: Merit fund allocations are considered augmentations to a program's core funding level and are subject only to the normal terms and conditions that apply to all funds used in supporting a program's core activities.

Merit Pool Allocations
by Rating Categories

	Merit Pool Allocation for Each Rating Category		
Merit Funding Rating Category	**Minimum Allocation**	**Residual Share**	**Program Merit Funding**
Category #1	Min(Cat. #1) = Total Merit Pool / n	2x	Min(Cat #1) + 2x
Category #2	Min(Cat. #2) = 70% * Min(Cat. #1)	x	Min(Cat #2) + x
Category #3	Min(Cat. #3) = 40% * Min(Cat. #1)	Zero	Min(Cat #3)
Category #4	Zero	Zero	Zero

where:

n	Number of programs evaluated (8 currently ... 29 after 4 years)
x	Residual Share = ("Total Merit Pool" minus "Total of Minimum Allocations") / #Shares (where #Shares = 2 times number of programs in "Category #1" plus the number of programs in "Category #2")

Merit Pool Allocations
Examples of Merit Funding Calculations for Hypothetical Rating Distributions

Rating Distribution (# of Programs)	Merit Funding Rating	Merit Pool Allocation for Each Rating: Minimum	From Residual	Merit Funding
4	"Category 1"	$ 100,000	$ 40,000	$ 140,000
4	"Category 2"	$ 70,000	$ 20,000	$ 90,000
2	"Category 3"	$ 40,000	$ -	$ 40,000
0	"Category 4"	$ -	$ -	$ -
10				

Total Merit Pool =	$ 1,000,000		
Total Minimum Allocations =	$ 760,000	Total -- Category 1	$ 560,000
Total Residual Allocations =	$ 240,000	Total -- Category 2	$ 360,000
Shares of Residual	12	Total -- Category 3	$ 80,000
Residual/Share	$ 20,000	Total Merit Pool	$ 1,000,000

Rating Distribution (# of Programs)	Merit Funding Rating	Merit Pool Allocation for Each Rating: Minimum	From Residual	Merit Funding
2	"Category 1"	$ 100,000	$ 90,000	$ 190,000
4	"Category 2"	$ 70,000	$ 45,000	$ 115,000
4	"Category 3"	$ 40,000	$ -	$ 40,000
0	"Category 4"	$ -	$ -	$ -
10				

Total Merit Pool =	$ 1,000,000		
Total Minimum Allocations =	$ 640,000	Total -- Category 1	$ 380,000
Total Residual Allocations =	$ 360,000	Total -- Category 2	$ 460,000
Shares of Residual	8	Total -- Category 3	$ 160,000
Residual/Share	$ 45,000	Total Merit Pool	$ 1,000,000

Rating Distribution (# of Programs)	Merit Funding Rating	Merit Pool Allocation for Each Rating: Minimum	From Residual	Merit Funding
4	"Category 1"	$ 100,000	$ 60,000	$ 160,000
2	"Category 2"	$ 70,000	$ 30,000	$ 100,000
4	"Category 3"	$ 40,000	$ -	$ 40,000
0	"Category 4"	$ -	$ -	$ -
10				

Total Merit Pool =	$ 1,000,000		
Total Minimum Allocations =	$ 700,000	Total -- Category 1	$ 640,000
Total Residual Allocations =	$ 300,000	Total -- Category 2	$ 200,000
Shares of Residual	10	Total -- Category 3	$ 160,000
Residual/Share	$ 30,000	Total Merit Pool	$ 1,000,000

Merit Pool Allocation

An Example of Merit Funding over Four Year Transition for Hypothetical Rating Distributions

(Year 1 and Year 2)

Year 1: (8 Programs)

Rating Distribution (# of Programs)	Merit Funding Rating	Merit Pool Allocation for Each Rating: Minimum	From Residual	Merit Funding
3	"Category 1"	$ 100,000	$ 70,000	$ 170,000
2	"Category 2"	$ 70,000	$ 35,000	$ 105,000
2	"Category 3"	$ 40,000	$ -	$ 40,000
1	"Category 4"	$ -	$ -	$ -
8				

Total Merit Pool =	$ 800,000		
Total Minimum Allocations =	$ 520,000	Total -- Category 1	$ 510,000
Total Residual Allocations =	$ 280,000	Total -- Category 2	$ 210,000
Shares of Residual	8	Total -- Category 3	$ 80,000
Residual/Share	$ 35,000	Total Merit Pool	$ 800,000

Year 2: (15 Programs)

Prior Distribution	New Group Distribution	Rating Distribution (# of Programs)	Merit Funding Rating	Merit Pool Allocation for Each Rating: Minimum	From Residual	Merit Funding
3	1	4	"Category 1"	$ 100,000	$ 80,000	$ 180,000
2	4	6	"Category 2"	$ 70,000	$ 40,000	$ 110,000
2	1	3	"Category 3"	$ 40,000	$ -	$ 40,000
1	1	2	"Category 4"	$ -	$ -	$ -
8	7	15				

Total Merit Pool =	$ 1,500,000		
Total Minimum Allocations =	$ 940,000	Total -- Category 1	$ 720,000
Total Residual Allocations =	$ 560,000	Total -- Category 2	$ 660,000
Shares of Residual	14	Total -- Category 3	$ 120,000
Residual/Share	$ 40,000	Total Merit Pool	$ 1,500,000

Merit Pool Allocation
An Example of Merit Funding over Four Year Transition for Hypothetical Rating Distributions
(Year 3 and Year 4)

Year 3: (22 Programs)

Prior Distribution	New Group Distribution	Rating Distribution (# of Programs)	Merit Funding Rating	Merit Pool Allocation for Each Rating: Minimum	From Residual	Merit Funding
4	4	8	"Category 1"	$ 100,000	$ 52,000	$ 152,000
6	3	9	"Category 2"	$ 70,000	$ 26,000	$ 96,000
3	0	3	"Category 3"	$ 40,000	$ -	$ 40,000
2	0	2	"Category 4"	$ -	$ -	$ -
15	7	22				
		Total Merit Pool =	$ 2,200,000			
		Total Minimum Allocations =	$ 1,550,000		Total -- Category 1	$ 1,216,000
		Total Residual Allocations =	$ 650,000		Total -- Category 2	$ 864,000
		Shares of Residual	25		Total -- Category 3	$ 120,000
		Residual/Share	$ 26,000		Total Merit Pool	$ 2,200,000

Year 4: (29 Programs)

Prior Distribution	New Group Distribution	Rating Distribution (# of Programs)	Merit Funding Rating	Merit Pool Allocation for Each Rating: Minimum	From Residual	Merit Funding
8	5	13	"Category 1"	$ 100,000	$ 44,000	$ 144,000
9	0	9	"Category 2"	$ 70,000	$ 22,000	$ 92,000
3	2	5	"Category 3"	$ 40,000	$ -	$ 40,000
2	0	2	"Category 4"	$ -	$ -	$ -
22	7	29				
		Total Merit Pool =	$ 2,900,000			
		Total Minimum Allocations =	$ 2,130,000		Total -- Category 1	$ 1,872,000
		Total Residual Allocations =	$ 770,000		Total -- Category 2	$ 828,000
		Shares of Residual	35		Total -- Category 3	$ 200,000
		Residual/Share	$ 22,000		Total Merit Pool	$ 2,900,000

Year 5 and after: (29 Programs)

Starting in year five, after all 29 programs have been phased in, this process continues in a similar manner. The overall merit rating distribution for 29 programs in year five is determined by replacing the year-one ratings of the first group of programs with their newer ratings set in year five. In year six, the ratings of the second group of programs from year two are replaced by their newer ratings set in year six, and so on for successive years.

Appendix E

Revised Policy Memorandum on NSGO Final Evaluation and Merit Funding (2005); April 8, 2005

[Stamped Date on Original: April 8, 2005]

MEMORANDUM FOR: Sea Grant Directors

FROM: Ronald C. Baird
Director

SUBJECT: Revised Policy Memorandum on NSGO Final Evaluation and Merit Funding (2005)

This policy document revises the previous, "Policy Memorandum on NSGO Final Evaluation and Merit Funding," (1999). Revisions incorporate recommendations from the Toll Committee Report (*"Review and Recommendations: Sea Grant Program Evaluation Process Report of the Sea Grant Review Panel's Program Evaluation Committee,"* October 2001) and the new requirements under the "National Sea Grant College Program Act Amendments of 2002" (Public Law 107–299). The policy was also reviewed for consistency with the memorandum, "Policy for the Allocation of Funds, FY2003 and Beyond," (March, 2003).

INTRODUCTION

To increase the overall performance and effectiveness of the Sea Grant network and associated institutional programs, the National Sea Grant Office (NSGO) tracks and analyzes the performance of each Sea Grant program. In 1998, the NSGO moved from the prospective evaluation of Sea Grant program proposals and their individual projects to a retrospective evaluation of overall program performance and accomplishment. The intent of evaluation is to improve Sea Grant's overall performance while providing incentives for strong performance. This change was first recommended by the Ocean Studies Board of the National Research Council in 1994 and subsequently endorsed by the institutions that comprise the Sea Grant network. The National Sea Grant College Program Act of 1998 (P.L. 105-160) codified this change and charged the Director of the National Sea Grant College Program to:

> *"204(3) With respect to sea grant colleges and sea grant institutes, the Director shall—*
> *"(A) evaluate the programs of sea grant colleges and sea grant institutes, using the priorities, guidelines, and qualifications established by the Secretary;*
> *"(B) subject to the availability of appropriations, allocate funding among sea grant colleges and sea grant institutes so as to—*
> *"(i) promote healthy competition among sea grant colleges and institutes;*
> *"(ii) encourage successful implementation of sea grant programs; and*
> *"(iii) to the maximum extent consistent with other provisions of this Act, provide a stable base of funding for sea grant colleges and institutes; and*
> *"(C) ensure compliance with the guidelines for merit review...."*

SEA GRANT PROGRAM EVALUATION

In response to the 1994 National Research Council/Ocean Studies Board Report and the 1998 Sea Grant reauthorization legislation, the NSGO introduced a system of performance-based reviews ("*Implementation of Program Evaluation Procedures in the National Sea Grant College Program,*" April 20, 1998) that continue to the present. Among other things, this requires

(1) Program Assessment Team Evaluations and (2) NSGO Final Evaluation Reviews.

1. Program Assessment Team Evaluations

An onsite evaluation by a Program Assessment Team (PAT) of each institution responsible for administering a Sea Grant program is conducted under the auspices of the Sea Grant Review Panel. The NSGO Director will notify university officials of the upcoming PAT. The role of the PAT is to assess the performance of a Sea Grant program with respect to a standard set of evaluation criteria and benchmarks and to make recommendations for the improvement of the program. The PAT Report and recommendations are used primarily to improve individual program performance and also to provide a basis for comparison among programs over the long term.

During the first cycle of PAT reviews (1998–2001), teams assigned a grade to each of the four major benchmark categories and an overall grade using appropriate weights:

Organizing and Managing the Program (20%)
Connecting Sea Grant with Users (20%)
Effective & Aggressive Long-Range Planning (10%)
Producing Significant Results (50%)

As a result of the Toll Committee Report, and subsequent recommendations by the Sea Grant Review Panel, the grading regime was changed for the second PAT cycle that began in 2003. Instead of marks for each of the four benchmark categories, the PAT will now provide a rating for each of 14 finer scale sub-elements under the four major benchmark categories and no overall grade. Those sub-elements and the weights assigned to each are listed below and described in detail in the PAT Manual:

Organizing and Managing the Program (20%)
Leadership of the Program (6%)
Institutional Setting and Support (4%)
Project Selection (2%)
Recruiting Talent (3%)
Effective and Integrated Program Components (5%)

Connecting Sea Grant with Users (20%)
Engagement with Appropriate User Communities (15%)
Partnerships (5%)

Effective & Aggressive Long-Range Planning (10%)
Strategic Planning Process (4%)
Strategic Plan Quality (4%)
Implementation Plan (2%)

Producing Significant Results (50%)
Contributions to Science and Technology (10%)
Contributions to Extension, Communications and Education (10%)
Impact on Society, the Economy and the Environment (25%)
Success in Achieving Planned Program Outcomes (5%)

Each sub-element will be given one of four possible ratings by the PAT:

- **Needs Improvement**—In general, performance does not reach the benchmark for this sub-element.
- **Meets Benchmark**—In general, performance meets, but does not exceed, the benchmark for this sub-element.
- **Exceeds Benchmark**—In general, performance goes beyond what would be required to simply meet the benchmark for this sub-element.
- **Highest Performance**—Performance goes well beyond the benchmark for this sub-element and is outstanding in all areas.

The PAT will provide a briefing for the Sea Grant Director and appropriate university officials at the end of the PAT visit. The ratings are presented at the debriefing and a rating sheet is also provided for the record. Following the PAT review, the chair of the assessment team provides a written PAT Report to the institution. The Sea Grant institution is encouraged to provide the NSGO a written response to the PAT Report. The comprehensive PAT Report and the institutional response to the report will become part of the record for the institutional program and both will be considered at the NSGO Review. Actions taken after the PAT by a Sea Grant institution in response to PAT recommendations will be acknowledged, but will not become a factor in the current NSGO ratings. All improvements made by the Sea Grant institution after the PAT will be more properly considered in the next PAT cycle.

2. NSGO Final Evaluation Reviews

The four-year NSGO Final Evaluation Review (henceforth, NSGO Review) is conducted by the NSGO in the year following the program's PAT visit (usually February). The evaluation relies primarily on the information provided by the program to the PAT, the PAT Report and ratings, and the institutional response to the PAT Report. A NSGO Final Evaluation Report (henceforth, NSGO Report) summarizes the findings of the NSGO performance review for that Sea Grant program over the last four-year review cycle. In addition to the report, the NSGO provides a performance rating to each Sea Grant program as part of the evaluation.

The primary objective of the NSGO Review is to provide local management with an assessment of performance and specific recommendations directed toward improvement and maintenance of existing program strengths. The second objective is to assign programs to a rating category that can be used in the allocation of a partial amount of Sea Grant funds. This rating was the basis during the first cycle of reviews (1998–2002) for allocating merit funding from a $3,000,000 pool of funds set aside in the Sea Grant budget for that purpose.

The seven or eight Sea Grant programs that were evaluated by a PAT in the prior calendar year are considered as a group and scheduled for NSGO Review every four years. The NSGO conducts the final evaluation during a one-week period, typically in the month of February. The criteria and benchmarks used in the NSGO Review are identical to those used by the PAT. Effort is taken to assure that all programs are evaluated in a similar manner using the same standard criteria and performance benchmarks listed above and described in detail in the PAT Manual.

The NSGO Director has mandated that all NSGO technical staff participate and be present for the entire review. One or more members of the National Sea Grant Review Panel, usually from the panel's executive committee, also attend this meeting as observers, which is consistent with the panel's oversight responsibilities for the conduct of program evaluation.

<u>Performance Information Considered</u>

In preparation for the NSGO Review, the NSGO Program Officer prepares materials for distribution. The PAT Report and the SG program's response provide the primary input to the NSGO review process.

Documents and reports are distributed to NSGO staff for study prior to the review, which include:

- The overview section from the program's briefing book prepared by the Sea Grant program for its PAT Review. This section includes a program description, the response to previous PAT recommendations, and a description of program accomplishments and impacts. (See *NSGO Guidelines for Program Assessment Briefing Books*, May 9, 2003)
- The Program Assessment Team Report as signed off by the PAT Chair
- The Sea Grant program's formal written response to the PAT Report
- The program's strategic and implementation plans

Collectively, the NSGO staff also has access to documents on file for each program, part of the continuous and ongoing communications that occur between a Sea Grant program and the NSGO. Some of these materials are less generally available to the PAT and represent additional information for the NSGO to use in the evaluation process. This includes:

- Annual progress reports
- Omnibus proposals
- Publications
- Archived information on accomplishments
- Trip reports and peer review panel visits by the Program Officer
- Topical Assessment Team reports (if any)
- Detailed Sea Grant funding information
- Supportive material deemed to be relevant by the Program Officer or staff

Structure of the NSGO Review

The NSGO Executive Director is responsible for planning the review and for the staff preparation needed to carry out the review. NSGO Program Officers are responsible for preparing a presentation on the programs that will be reviewed. The NSGO Director facilitates the evaluation sessions during review week. Each half-day session is focused on a single program and is reviewed on its own merits and not in direct comparison with other programs.

Since Program Officers play a central role in the NSGO Review, it is the NSGO's policy not to reassign Program Officers in mid-cycle, if at all possible. Assignments are made with the goal of maintaining continuing associations between the federal program officer and a Sea Grant program over the review cycle, or longer. However, at times this will not be feasible due to NSGO staff turnover.

To begin the review of each program, the NSGO program officer provides an overview of the program's performance since the last PAT (five years for the second cycle of reviews, but normally four years). Each program officer follows a prescribed format using a standardized presentation template that ensures consistency of the kinds of information being presented. The template follows the benchmarks and indicators of performance from the PAT Manual. Evaluation-related materials on file in the NSGO (see above) are considered where appropriate. For example, participation in national competitions and responsiveness to network-wide activities have bearing on successful performance, and often the NSGO will have a better perspective here than would the PAT.

Following the Program Officer's presentation, the NSGO director facilitates a discussion of the program. The review is structured to consider the same criteria and benchmarks addressed by the PAT. Each of the four major evaluation criteria and the 14 sub-elements are discussed in succession, including the PAT findings and ratings. All PAT recommendations are reviewed. Those deemed most critical from the NSGO perspective are highlighted for inclusion in the NSGO Report. Where appropriate, PAT recommendations may be modified and additional recommendations developed based on the NSGO Review. During the review process, "best management practices" are identified for subsequent promulgation to the Sea Grant network.

The discussions and findings from the NSGO Review form the basis for a report that is prepared under the direction of and signed by the NSGO Director. The NSGO Report is best understood when read in conjunction with the PAT Report, which will be included with the NSGO Report when distributed. It is the NSGO's intention to complete the report and transmit it to the Sea Grant Director within 30-days of the end of the NSGO Review. While the NSGO Report findings and ratings are considered final, the draft report will be sent to the program director for factual review and correction of minor errors prior to final distribution. A ten-day turnaround period is considered a reasonable time for directors to respond, but extensions may be requested. The NSGO report will be sent only to the Sea Grant program. The Sea Grant program director can decide how to use the NSGO Report within their university. The NSGO

Report for a given program will be distributed to the Panel members who served on that particular PAT.

Under certain circumstances, the NSGO director may elect to send a special letter of findings to the institution. These specials reports, it is expected, will be used infrequently and only in highly unusual cases that warrant communication at a higher administrative level in the university.

NSGO RATING DECISIONS

One objective of the NSGO Review process is to provide a consistent approach to rating Sea Grant programs. The intensive, weeklong PAT evaluation by a team of experts, who interact with university officials, constituents, and government officials, provides credible information from which to judge a program's performance. The NSGO Review provides an additional assessment of performance that adds to the PAT review in several salient ways:

- Performance-relevant information available to the NSGO results, not only from the PAT process, but also from a continuous process of evaluation and dialogue between the NSGO and the Sea Grant program over the full four-year cycle.

- For the NSGO Review, the institution's formal response to the PAT's findings and recommendations is available and explicitly considered. This additional input is critical information for the NSGO Review and can often provide clarifying information on program performance.

- The NSGO Review provides a broad perspective across seven to eight programs each year, and across all Sea Grant programs over a four-year cycle. While the NSGO evaluates programs individually, by considering a group of programs at the same time and with the same reviewers, more consistency for assigning ratings can be achieved.

The rating of a program involves the use of judgment in weighing the qualitative and quantitative evidence available. Following extensive discussion of a program's performance under each evaluation criteria, the NSGO staff provides their individual rating ranging from 1(highest) to 4 (lowest) for the 14 evaluation sub-elements. The NSGO director will set a minimum level of experience that will be required of new NSGO staff members before they will be asked to contribute their individual rating of programs. Nonetheless, new technical staff members are expected to be present and to participate in the discussions.

The NSGO rating for a program is derived from reaching a broad consensus (two-thirds majority) of individual NSGO staff ratings for the 14 sub-elements using the PAT results as reference. Taking into account the proportional weighting of each of the 14 sub-elements (e.g., Contributions to Science and Technology - 10%) and the NSGO consensus ratings for each, a program score is calculated. The NSGO final rating for the program is determined by locating a program's score along a fixed four-category rating scale for merit funding and a variable two-category rating scale for bonus funding. Merit funding and bonus funding allocations are discussed in detail below.

The NSGO Review is a semi-autonomous review that significantly weighs and is informed by the PAT findings and ratings. As would be expected, the findings and ratings of the PAT and the NSGO are in agreement in the large majority of cases. As a matter of policy, however, if there is not a broad consensus agreement (two-thirds majority) on a particular sub-element rating, the NSGO assigns a rating consistent with the PAT rating for that sub-element.

Final ratings for the group of seven or eight programs are considered at the last session of the review week. The NSGO final ratings are reviewed and considered for adjustment, if NSGO staff offers a convincing case for reconsideration. Rating adjustments result, as for all NSGO ratings, only from a broad consensus agreement (two-thirds majority). All decisions to change a PAT rating are ultimately the final responsibility of the NSGO Director. At the conclusion of the session, all NSGO ratings are considered final.

The final NSGO ratings are used to assign each program to a merit-funding category and are also interleaved with the last rating of all other Sea Grant programs to determine eligibility for bonus-funding categories.

NSGO RATINGS AND MERIT FUND ALLOCATIONS

First Cycle (1998/99–2002/03)

In the first cycle of merit funding, the NSGO established a pool of funds in the Sea Grant budget to be allocated to individual Sea Grant programs on the basis of overall performance. The NSGO developed systematic procedures to rate each Sea Grant program for the purpose of allocating funds from this merit pool. During Cycle 2, merit fund allocations will continue to be made in a manner similar to Cycle 1, with slight modifications.

Merit Funding

Sea Grant programs that have reached institutional or college status are assigned to one of four merit categories. Categories 1 and 2 are reserved for programs that achieve the highest levels of performance. Category 3 denotes programs meeting performance benchmarks, while programs assigned to Category 4 have significant deficiencies. Programs assigned to categories 1, 2 and 3 qualify for merit pool allocations over the next four years. If a program fails to meet 20 percent or more of the weighted benchmarks, it will be considered as having "significant deficiencies" and assigned to Category 4. Programs assigned to Category 4 do not receive a merit pool allocation during the four-year period.

The merit pool allocation consists of two parts:

- A minimum allocation that is fixed for four-years (assuming level funding), and
- A residual share component that is variable and may change each year depending upon the performance ratings of all programs that have been reviewed.

Added together, these two components determine each program's merit funding allocation for a given year.

The minimum allocation is a fixed percentage of the merit pool that a program can expect to receive over the course of the next four years. Assuming level funding of the merit pool, this amount will remain the same each year. The fixed minimum component for a program in Category 1 is calculated by dividing the total amount of funds in the merit pool by the number of programs (e.g. $3 million merit pool ÷ 30 programs = $100,000 per program in Category 1). A program in Category 2 and Category 3 receives a minimum allocation of 70 percent and 40 percent respectively of that received by a program in Category 1. In the above example of a $100,000 minimum allocation for Category 1 programs, the minimum component for programs in Category 2 and Category 3 would be $70,000 and $40,000 respectively. A program assigned Category 4, "significant deficiencies," would not receive a merit funding allocation. Once the amount in the merit pool is determined, the fixed minimum component remains unchanged until the total merit pool amount changes.

The residual share of merit funding depends upon the distribution of ratings across all programs for a given year. The amount of the merit pool that remains unallocated after meeting the minimum allocations, the "residual" amount, is distributed to programs in Category 1 and Category 2 only. Category 1 programs gets twice as much of the residual as those in Category 2 programs. Category 3 programs do not receive a share of the residual. A new residual share is calculated every year. The NSGO Director could cap an award if the residual share exceeds 10 percent of the merit pool, although in practice this is unlikely to occur.

Merit Funding

- Category 1 = (Merit Pool ÷ 30) + 2 shares of residual
- Category 2 = 70% of (Merit Pool ÷ 30) + 1 share of residual
- Category 3 = 40% of (Merit Pool ÷ 30) + no share of residual
- Category 4 = no merit funding

Following the NSGO Review each year, the new ratings for the seven or eight programs replace their prior rating and the merit pool allocations are recalculated. Calculating the allocations each year assures that all programs have the same merit funding opportunities, regardless of the year they are reviewed. The distribution of the $3 million merit pool at the end of the first cycle is shown below.

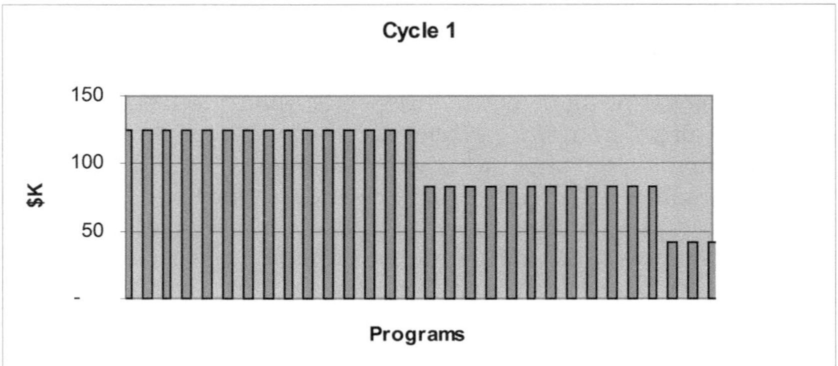

Rating	Merit Funding	Merit Pool Allocation for Each Rating		
Distribution (# of Programs)	Rating	Minimum	From Residual	Merit Funding
15	"Category 1"	$ 100,000	$ 25,714	$ 125,700
12	"Category 2"	$ 70,000	$ 12,857	$ 82,900
3	"Category 3"	$ 40,000	$ -	$ 40,000
0	"Category 4"	$ -	$ -	$ -
30				

Total Merit Pool =	$ 3,000,000		
Total Minimum Allocations =	$ 2,460,000	Total Category 1	$ 1,885,500
Total Residual Allocations =	$ 540,000	Total Category 2	$ 994,800
Shares of Residual	42	Total Category 3	$ 120,000
Residual/Share	$ 12,857	Total Merit Pool	$ 3,000,300

Second Cycle (2003/04–2006/07)

Sea Grant Reauthorization Legislation (2002)

New provisions of the "National Sea Grant College Program Act Amendments of 2002" (Public Law 107–299) impose new requirements for evaluation of Sea Grant college and institutional programs. The law now requires the Director of the National Sea Grant College Program to rate such programs according to their relative performance into at least five categories, with each of the two best-performing categories containing at most 25 percent of the programs. In particular,

Section 3, Requirements Applicable to National Sea Grant College Program, states:

(b) PROGRAM EVALUATION AND RATING.—

(1) EVALUATION AND RATING REQUIREMENT.—Section 204(d)(3)(A) of the National Sea Grant College Program Act (33 U.S.C. 1123(d)(3)(A)) is amended to read as follows:

"(A)(I) evaluate the performance of the programs of sea grant colleges and sea grant institutes, using the priorities, guidelines, and qualifications established by the Secretary under subsection c), and determine which of the programs are the best managed and carry out the highest quality research, education, extension, and training activities; and

"(ii) rate the programs according to their relative performance (as determined under clause (I)) into no less than 5 categories, with each of the 2 best-performing categories containing no more than 25 percent of the programs."

Public Law 107–299 also requires the Secretary of Commerce to distribute all appropriations in excess of FY2003 levels to any combination of: (1) Sea Grant programs, according to their performance rating; (2) national strategic investments; (3) Sea Grant program qualifying activities; and (4) Sea Grant colleges or institutes designated after this Act's enactment, but not yet evaluated.

Section 7, Authorization of Appropriations, states:

c) DISTRIBUTION OF FUNDS.—In any fiscal year in which the appropriations made under subsection (a)(1) exceed the amounts appropriated for fiscal year 2003 for the purposes described in such subsection, the Secretary shall distribute any excess amounts (except amounts used for the administration of the sea grant program) to any combination of the following:

"(1) sea grant programs, according to their rating under section 204(d)(3)(A);

"(2) national strategic investments authorized under section 204(b)(4);

"(3) a college, university, institution, association, or alliance for activities that are necessary for it to be designated as a sea grant college or sea grant institute;

"(4) a sea grant college or sea grant institute designated after the date of enactment of the National Sea Grant College Program Act Amendments of 2002 but not yet evaluated under section 204(d)(3)(A)."

Merit Funding and Bonus Funding Combined

In summary, the three key provisions of Public Law 107–299 that will affect the ratings and allocation of funds during the second cycle of reviews are:

- The NSGO is required to rate programs according to their relative performance and assign programs into no less than five categories.
- Each of the top two categories cannot contain more than 25 percent of the Sea Grant programs.
- Appropriations above the FY2003 level can be allocated according to these ratings.

In order to meet these requirements, the NSGO will adopt a two-tier approach to funding allocations related to performance evaluations.

- The first tier, or "merit funding" tier, retains the framework of the Cycle 1 merit funding. All programs will continue to be assigned to a merit-funding category. Programs assigned a rating of Category 1 (highest), 2 or 3, based on the NSGO Review, will receive merit funding allocated similarly to the Cycle 1 allocation procedures. Categories 1 and 2 are reserved for programs that achieve the highest levels of performance. Category 3 denotes programs that meet performance benchmarks.

- The second tier, or "bonus funding" tier, will at times be used to allocate part or all of the funds appropriated in excess of the FY2003 appropriation. Bonus funding would go only to programs that are rated in Category 1 (best-performing category) and are rated among the top programs in "Category 1" (each of the 2 best-performing categories containing no more than 25 percent of the programs). Currently, this would allow up to 14 programs to receive bonus funding, or up to seven programs in each of the two bonus funding categories. It would be expected that the NSGO would maintain a 2:1 ratio in the "bonus pool" for the two "bonus" groups in Category 1.

In combination, the three merit categories plus the two bonus categories give the five groups mandated by Congress:

First group (1A)—Top "Category 1" programs (currently limited to 7) receive "Category 1" merit funds + the higher bonus

Second group (1B)—Next "Category 1" programs (currently limited to 7) receive "Category 1" merit funds + the smaller bonus

Third Group (1C)—All other "Category 1" programs (no limit) receive "Category 1" merit funds / no bonus

Fourth Group—all "Category 2" programs receive "Category 2" merit funds / no bonus

Fifth Group—all "Category 3" programs receive "Category 3" merit funds / no bonus

Each program is assigned to a merit-funding category (Category 1, 2 or 3) that will not change over the four-year period. There is no interim grading of programs in the "outyears". Programs are evaluated and rated once every four years through the PAT and NSGO process, and the program's rating is in effect for the full four years. The one change that may occur over time is a program's relative position in the new rating categories mandated by Congress, or in Sea Grant terminology, the two new bonus categories.

Assignment to the two new bonus categories is dependent not only on a program's rating, but also on the distribution of the ratings of all programs. Consequently, with respect to the bonus funding only, it is possible for a Sea Grant program not being reviewed to be affected. The ratings of the seven or eight programs reviewed yearly may reorder the distribution of ratings across programs. Each year it is possible for a program not reviewed to move into or out of a bonus category (e.g., from Group 1B to 1C or vice versa) or to move up or down between the two bonus categories (e.g., from Group 1A to 1B or vice versa).

Each year, the NSGO Director may add funds to the merit pool and/or the bonus pool in response to actual appropriations. In the future, were appropriations to increase substantially, increases in merit funding would be a primary mechanism for maintaining and enhancing Sea Grant's enabling infrastructure. Currently, as many as 14 programs in Category 1 would receive bonus funding in addition to merit, in any given year. However, because Congress limits the number of bonus programs, the actual allocations would depend on the distribution of the merit ratings and the number of "Category 1" programs. As such, the merit pool would be expected to be larger relative to the bonus pool.

The NSGO expects to provide preliminary notice to programs of the next fiscal year's merit funding and changes in bonus categories, if any, following finalization of the NSGO Review. A funding letter will be prepared that will go to each Sea Grant program to indicate the dollar level of the merit funding allocation a program will receive in the following fiscal year, assuming level funding. The letter will also indicate whether a program's rating makes it eligible for either of the two bonus categories, along with a preliminary estimate of bonus funding for the following fiscal year, again assuming level funding. Normally, this letter will be sent 10 months or more ahead of the anniversary dates for renewal of omnibus grant awards.

The hypothetical example below shows how a $3 million merit funding pool plus a $1 million bonus funding pool might be allocated assuming 20 programs have been rated in "Category 1".

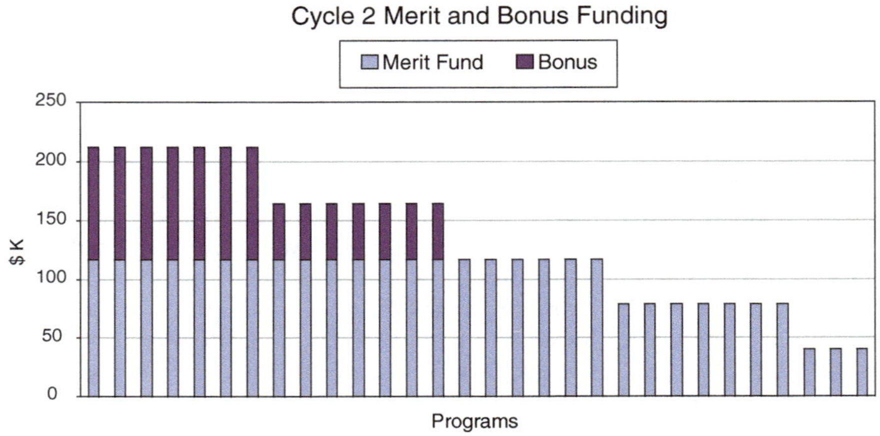

It should be noted that it is possible that a particular group would not have any programs assigned to it. For example, if there were 14 or fewer Category 1 programs, the Third Group (1C) would have no programs assigned to it.

Programs with Significant Deficiencies: The major goal of the evaluation process is to help programs improve. If as a result of the NSGO Review, a program is determined to have a significant number of deficient program elements (fails to meet 20 percent or more of the weighted benchmarks), the program will be assigned to the "Significant Deficiencies" (Category 4) and would not be eligible to receive merit funds over the next four-year cycle.

While occurring very rarely, if the NSGO Review determines that a program should receive a rating of "Significant Deficiencies," a corrective action plan will be required to address all the deficient elements. The action plan identifies any changes in goals, organization, procedures, planning, and operations that need to be implemented to correct the deficiency. The action plan is a joint effort of the Sea Grant institution and the NSGO. The plan should be in place within six months of notification to the Sea Grant program of the "Significant Deficiencies" (Category 4) rating and the need of corrective action. Failure to fully implement a corrective action plan and to show significant improvement by the two-year mid-cycle mark, as determined by an assessment team, could result in a program having its core funding reduced or decertification of Sea Grant college or institutional status.

In addition, the NSGO Director may also require a program rated in categories 1, 2 or 3 to submit a corrective action plan for a particular area of the program. If the NSGO Review finds that a program fails to meet the benchmark for a sub-element, the requirement for a partial corrective action plan will be identified in the NSGO Report.

Appendix F

A Multivariate Analysis of Potential Biases in the Final Evaluation Scores

Because bivariate relationships can be obscured if the data generating processes are multivariate, the data were also examined using a multivariate regression approach. As was also true in the case of the bivariate statistical analyses, the multivariate model was designed to explore the statistical significance of potential sources of bias in the determination of National Sea Grant Office Final Evaluation Review (FE) scores. Thus the model did not include measures of program accomplishments and success, but instead assumed that the Program Assessment Team (PAT) and FE scores provide accurate assessments of program quality according to the assessment criteria, but might be subject to random errors associated with differences between Cycle 1 and Cycle 2, the number of years that particular NSGO program officers are associated with: particular Sea Grant programs; program seniority; the size of state and federal budget allocations awarded to programs; the within cycle order of review of programs; and the number of years that particular program officers have served as program officers. The general linear model that was estimated can be represented by:

$$FE_{ij} = f\left[\begin{array}{l} Cycle_j,\ PO\ Continuity_{ij},\ Program\ Maturity_{ij},\ State\ Budget_{ij}, \\ Federal\ Budget_{ij},\ Order\ of\ Review_i,\ PO\ Seniority_{ij} \end{array}\right]$$

where *Cyclej* is a binary variable used to differentiate between scores awarded in Cycle 1 and Cycle 2; *PO Continuity* is the number of years that a particular NSGO program officer is assigned to the ith individual Sea

Grant program during the jth review cycle; *Program Maturity* is the number of years that elapsed between the initial chartering of the ith individual Sea Grant program and the jth review cycle; *State Budget* is the average state budget allocated to the ith individual Sea Grant program for 2000 through 2002 for observations from Cycle 1 and the 2003 budget for Cycle 2; *Federal Budget* is the average federal budget allocated to the ith individual Sea Grant program for 2000 through 2002 for observations from Cycle 1 and the 2003 budget for Cycle 2; *Order of Review* is a pair of binary variables used to differentiate between individual Sea Grant programs reviewed in the first or second year of each cycle from those that were reviewed in the third or fourth year of that cycle; and *PO Seniority* is a set of binary variables used to differentiate between individual Sea Grant programs that were reviewed by program officers with one or less, 2 or 3, 4 to 10, or more than 10 years of experience as program officers. With observations from Cycle 1 and Cycle 2, there were 44 observations available to use in the analysis. The initial model coefficient estimates are:

	Coefficients	*Standard Error*	*P-value*
Intercept	2.723	0.608	0.000
Cycle Dummy	0.068	0.156	0.667
PO Continuity	-0.087	0.038	0.029
Program Maturity	-0.040	0.019	0.046
State Budget	1.17E-07	2.32E-07	0.617
Federal Budget	7.68E-08	1.11E-07	0.493
Prog Reviewed in Year 1	0.049	0.186	0.793
Prog Reviewed in Year 2	0.088	0.162	0.591
PO Experience < or = 1 year	-0.277	0.411	0.505
PO Experience 2 to 3 years	0.093	0.212	0.664
PO Experience 4 to 10 years	-0.117	0.146	0.428

The structure of the model can be viewed as an attempt to explain variations in FE scores for the individual programs using information or proxy information for potential sources of bias that were suggested by the individual Sea Grant program directors. Thus, if the model were to provide accurate predictions of the FE scores, there would be evidence to support the concerns of the individual Sea Grant program directors. The value of R^2 (0.292) indicates that the estimated model accounts for 29.2 percent of the observed variation in FE scores. The F-statistic (1.359) is used to test whether the model estimates provide a statistically significant improvement over simply using the average of all FE scores as a predictor. The null hypothesis for the test is that the sum of squared deviations of the estimates is not significantly different from the sum of squared deviations about the mean. Because the probability that the null hypothesis is true (0.242) is greater than 5 percent, the null hypothesis cannot be rejected.

Although the overall model performance does not lend credence to the hypothesized biases, it is instructive to look at the model coefficients. The coefficients are the partial derivatives of the model with respect to the explanatory variables. That is, the coefficients are the estimated changes in the value of the FE score for a marginal increase in the associated explanatory variable, holding the value of all other explanatory variables constant.

The coefficient associated with the *Cycle* dummy suggests that there has been an average increase of 0.068 points in the scores of programs in Cycle 2 relative to the scores of programs in Cycle 1. This increase could be due to across-the-board degradation in the programs or tougher grading, but the difference could also have resulted from pure chance. Indeed, the probability that a value of 0.068 could have been observed even if the truth were that there is no effect is 0.667; consequently, it can be concluded that the estimated difference is not significantly different from zero.

The *PO Continuity* variable is associated with a coefficient of -0.087. This suggests that for each additional year that a particular program officer spends working with a particular Sea Grant program, the average FE score falls by 0.087 points. This is consistent with public testimony that suggested that the scores would be lower for individual Sea Grant programs that enjoyed longer working relationships with their program officers. Consequently, the relevant null (no effect) hypothesis is that this coefficient is not significantly greater than zero. Because the probability of observing an estimate of -0.087 if the true value of this coefficient were greater that or equal to zero is 0.014, the null hypothesis can be rejected. That is, there is statistical support for the assertion that individual Sea Grant programs with long-term relationships with their program officers scored lower than programs with less program officer continuity.

The coefficient associated with the *Program Maturity* variable (-0.040) suggests that for every additional year of age, program scores decline by 0.040 points. Because testimony suggested that there is an inverse relationship between program age and the FE score, the null hypothesis is that the estimated coefficient is greater than or equal to zero. Because the probability that we would observe an estimate of -0.040 if the true value of the coefficient were greater than or equal to zero is 0.023,[1] the null

---

[1]The p-value for a 1-tail test is one half the magnitude of the p-value for a 2-tail test; Excel's regression output defaults to a 2-tail p-value.

hypothesis can be rejected; there is statistical support for the assertion that mature programs are scored lower than newer programs.

The coefficients associated with the magnitude of state and federal budgets allocated to the individual Sea Grant programs indicate that programs with larger budgets earn higher scores, but the effect is miniscule: a $1 increase in the individual program's state budget is associated with an increase of 1.17E-07 in the score, and a $1 increase in the individual program's federal budget is associated with an increase of 7.68E-08 in the score. That is, to increase the score by 0.1 point, the individual program's state budget would need to be increased by about $8.5 million or the individual program's federal budget would need to be increased by about $13 million. Moreover, the standard errors of the coefficient estimates are so large that the probabilities that differences in the magnitude of state and federal budget allocations have no effect on FE scores are greater than 50 percent.

The effect of *Order of Review* is represented by two binary variables, so the influence of order of review must consider both coefficients together. The appropriate test is an F-test that compares the predictive ability of the model presented above and a model that differs from the above model by excluding the two binary variables used to represent the order of review. The probability that the order of review has no statistically significant influence on the FE score is 93 percent.

The effect of *PO Seniority* is represented by three binary variables, each of which represents the average difference in scores awarded to programs with the most senior program officers relative to the scores awarded to programs with one of the three categories of less experienced program officers. The statistical significance of the influence of program officer seniority is tested with an F-test similar to the test applied for *Order of Review*. The probability that program officer seniority has no statistically significant influence on the FE score is 64 percent.

Because preliminary analysis failed to eliminate the possibility that *PO Continuity* or *Program Maturity* exercise statistically significant influence on FE scores, the model was respecified using only those variables as explanations of the observed variation in final scores. The restricted model coefficient estimates are:

	Coefficients	*Standard Error*	*P-value*
Intercept	2.504	0.530	2.74E-05
PO Continuity	-0.079	0.029	0.010
Program Maturity	-0.040	0.019	0.046
State Budget	-0.023	0.015	0.137

Although the value of R^2 (0.184) for this simpler model is smaller than the R^2 for the initial model (0.292), the difference in model performance is not statistically significant.[2]

In the restricted model, the coefficient (-0.079) associated with the *PO Continuity* variable suggests that for each additional year that a particular program officer spends working with a particular individual Sea Grant program, the average FE score falls (is improved) by 0.079 points. Again, because public testimony suggested that the scores would be lower for programs that enjoyed longer working relationships with their program officers, the null (no effect) hypothesis is that this coefficient is not significantly greater than zero. Because the probability of observing an estimate of -0.079 if the true value of this coefficient were greater that or equal to zero is 0.005, the null hypothesis can be rejected. That is, there is again statistical support for the assertion that individual Sea Grant programs that have enjoyed long term relationships with their program officers scored lower (better) than programs with less program officer continuity.

The coefficient associated with the *Program Maturity* variable (-0.023) suggests that for every additional year of age, program scores decline by 0.023 points. Because testimony suggested that there is an inverse relationship between program age and the FE score, the null hypothesis is that the estimated coefficient is greater than or equal to zero. However, because there is a 0.069 probability of observing an estimate of -0.023 even if the true value of the coefficient were greater than or equal to zero, the null hypothesis cannot be rejected, thus there is insufficient statistical support for the assertion that mature programs are scored lower than newer programs.

The results of the restricted model suggest that the model could be further simplified without statistically significant loss of performance. The coefficient estimates for a simple linear regression model are:

---

[2]If the true difference in performance between the initial model and the restricted model were zero, the probability of observing this large of a decrease in model fit with the elimination of 8 explanatory variables is 0.747.

	Coefficients	*Standard Error*	*P-value*
Intercept	1.715	0.103	0.000
PO Continuity	-0.077	0.030	0.013

Although the value of R^2 (0.138) for this model is again smaller than the R^2 for the initial model (0.292), the difference in model performance is not statistically significant.[3]

In this model, the *PO Continuity* variable is associated with a coefficient of -0.077, suggesting that for each additional year that a particular program officer spends working with a particular individual Sea Grant program, the average FE score falls (improves) by 0.077 points. Again, because public testimony suggested that the scores would be lower for Sea Grant Colleges and Institutes that enjoyed longer working relationships with their program officers, the null (no effect) hypothesis is that this coefficient is not significantly greater than zero. Because the probability of observing an estimate of -0.077 if the true value of this coefficient were greater that or equal to zero is only 0.007, the null hypothesis can be rejected. That is, there is again statistical support for the assertion that individual Sea Grant programs with long term relationships with program officers are scored lower (better) than programs with less program officer continuity.

In summary, the results of the multivariate analysis are generally consistent with the results of the bivariate analyses and do not support the suggestions that the FE scores are biased as a result of program officer seniority, program funding levels, program maturity, order of review within a cycle, or between Cycle 1 and Cycle 2. However, there is persistent and statistically significant evidence that program officer continuity with the individual Sea Grant program is inversely related to the FE score. Indeed, there is less than a 0.007 probability of observing an estimate as large as $|-0.077|$ if the true value of the coefficient were zero.

The analysis suggests that knowing how long a program officer has been assigned to a state program carries information that is reflected in the FE scores, but the analysis does not identify whether the observed

[3]If the true difference in performance between the initial model and the restricted model were zero, the probability of observing this large of a decrease in model fit with the elimination of 9 explanatory variables is 0.622.

effect is a consequence of program officers representing the program during the PAT or FE or due to the program officers helping to mentor the individual Sea Grant programs or some other cause. While an effect of 0.077 points seems small, in 2004-05, the average difference between Category 1A and Category 1B was 0.13 points, the predicted equivalent magnitude of a two-year difference in the length of time that a particular program officer is assigned to a particular individual Sea Grant program. The average difference between Category 1B and Category 1C is of a similar magnitude. Thus for two otherwise identical individual Sea Grant programs that deserve to be rated in Category 1A—one with a new program officer and one with a program officer who has been with an individual Sea Grant program for 4 years—the program with the new officer would be expected to score 0.307 points higher (worse), a difference large enough to move it from Category 1A to Category 1C.

Appendix G

Expected Indicators of Performance and Other Issues of Importance

(Reprinted from NSGO, 2005a, pp. 57-59.)

There are many ways that information regarding the expected indicators of performance can be presented in the briefing book and during the PAT review. The following outline is intended only to provide an example of how this information might be organized in the briefing book appendix; it is anticipated that the presentation will be tailored to suit the needs of each individual program.

I. Organizing and Managing the Program Indicators:

A. Leadership of the Program

1. *Management Team composition and responsibilities (0.5 page narrative / organization chart)*
2. *Percentage time Director and staff devote to SG (FTEs)*
3. *Advisory Boards membership and function (expertise, meeting schedule, recommendations) (0.5 page)*

B. Institutional Setting and Support

1. *Setting of the program within the university or consortium organization and reporting structure (0.5 page wire diagram)*

C. Project Selection

1. *Brief description of the process used to develop RFP priorities (0.5 page)*
2. *Number of Preproposals and Full Proposals submitted, and institutions represented / institutions available in state*

	1st Biennial Cycle	2nd Biennial Cycle	*Total*
Preproposals	#	#	#
Full Proposals	#	#	#
Institutions	#	#	#

3. *Brief description of the review process including composition of review panels (0.5 page)*

D. Recruiting Talent

1. *New vs. continuing projects and PI's*
2. *Recruitment of PI's/institutions*
3. *Relative success of home institution*
4. *Success in national competitions*
5. *Regional/multi-program projects*

	Yr 1	Yr 2	Yr 3	Yr 4	Total
New projects	#	#	#	#	#
Continuing projects	#	#	#	#	#
New PIs	#	#	#	#	#
Success of home institution	#	#	#	#	#
New institutions	#	#	#	#	#
Success in national competitions	%	%	%	%	%
Regional & multi-program projects	#	#	#	#	#

E. Integrated Program Components

1. *Integration of outreach and research program elements (0.5 page)*
2. *Core Federal and matching funds and distribution among program elements over the last 8 years*

Year	SG	Match	Distribution (Research, Extension, Education, Communications, Program Development, Administration,)
Year 1	$	$	R= $ Ex=$ Ed=$ C=$ PD=$ A=$
Year 2	$	$	R= $ Ex=$ Ed=$ C=$ PD=$ A=$
Year 3	$	$	R= $ Ex=$ Ed=$ C=$ PD=$ A=$
Year 4	$	$	R= $ Ex=$ Ed=$ C=$ PD=$ A=$
Year 5	$	$	R= $ Ex=$ Ed=$ C=$ PD=$ A=$
Year 6	$	$	R= $ Ex=$ Ed=$ C=$ PD=$ A=$
Year 7	$	$	R= $ Ex=$ Ed=$ C=$ PD=$ A=$
Year 8	$	$	R= $ Ex=$ Ed=$ C=$ PD=$ A=$

3. *Leveraged funding from partners (NOAA, other Federal, State and local) for the program over the last 8 years*
4. *National competition funding (NSIs, pass through awards) over the last 8 years*
5. *Additional Program Funding through grants, contracts and development activities over the last 8 years*

II. Connecting Sea Grant with Users Indicators:

A. Engagement with Appropriate User Communities

1. *Leadership by staff on boards and committees (0.5 page)*

B. Partnerships (0.5 page)

1. *Effective local, regional and national interactions/collaborations including NOAA programs*

2. Leveraged funding from partners (NOAA, other Federal, State and local) for the program

III. Effective and Aggressive Long-Range Planning Indicators:

A. Strategic Planning Process (1 page)

1. *Stakeholder and staff involvement (numbers and duration) and integration of input into planning*
2. *Plan development (or reassessing priorities), selection process, and clear articulation of priorities*

B. Strategic Plan Quality (1 page)

1. *Short to long-term functional and management goals established*

C. Implementation Plan (1 page)

1. *Distribution of investment/effort to meet strategic plan priorities*
2. *Identification of short to long-term benchmarks*
3. *Work plan developed for integration of program elements*

IV. Producing Significant Results Indicators:

A. Contributions to Science and Technology (1 page)

1. *Number of publications (journal articles, book chapters, reports, etc.) (Publication list is a separate Appendix)*
2. *Invention disclosures and patents*

B. Contributions to Extension, Communications and Education

1. *Extension – Sponsorship of education programs and target audience participation; Internal evaluation processes for products and programs; Staff and product awards*
2. *Communications – Number, list and diversity of products produced (print, audio, video, web, etc.); Staff and product awards*
3. *Education – Numbers of graduate and undergraduate students supported, including fellowships and internships; Staff and product awards*

Students	Year 1	Year 2	Year 3	Year 4	Total
Undergraduate	#	#	#	#	#
Graduate	#	#	#	#	#

C. Impact on Society, the Economy, and the Environment—This element should make up the bulk of the discussion in Section 1 of the briefing book, but programs may wish to include additional information here.

1. Descriptions of the most important impacts
2. Positive environmental impacts and economic and social benefits resulting from changes in behavior of individuals, businesses, and institutions

D. Success in Achieving Planned Program Outcomes—There are no expected indicators for this element that need to be included in the appendix. Programs should address this element as part of Section 1 and/or during their PAT review.

Appendix H

U.S. Code, Title 33, Chapter 22

United States Code Provisions Affecting the National Sea Grant College Program

- United States Codes
  - TITLE 33 - NAVIGATION AND NAVIGABLE WATERS
    - CHAPTER 22 - SEA GRANT COLLEGES AND MARINE SCIENCE DEVELOPMENT
      - SUBCHAPTER II - NATIONAL SEA GRANT COLLEGE PROGRAM
        - Section 1121. Congressional Declaration Of Policy
        - Section 1122. Definitions
        - Section 1123. National Sea Grant College Program
        - Section 1124. Program Or Project Grants And Contracts
        - Section 1124a. Repealed. Pub. L. 105-160, Sec. 6, Mar. 6, 1998, 112 Stat. 24
        - Section 1125. Repealed. Pub. L. 102-186, Sec. 4(A), Dec. 4, 1991, 105 Stat. 1283
        - Section 1126. Sea Grant Colleges And Sea Grant Institutes
        - Section 1127. Fellowships
        - Section 1128. Sea Grant Review Panel
        - Section 1129. Interagency Cooperation
        - Section 1130. Repealed. Pub. L. 102-186, Sec. 5(A), Dec. 4, 1991, 105 Stat. 1283
        - Section 1131. Authorization Of Appropriations
        - Subchapter Notes

Section 1121. Congressional declaration of policy

(a) Findings

The Congress finds and declares the following:

(1) The national interest requires a strategy to -

(A) provide for the understanding and wise use of ocean, coastal, and Great Lakes resources and the environment;

(B) foster economic competitiveness;

(C) promote public stewardship and wise economic development of the coastal ocean and its margins, the Great Lakes, and the exclusive economic zone;

(D) encourage the development of forecast and analysis systems for coastal hazards;

(E) understand global environmental processes; and

(F) promote domestic and international cooperative solutions to ocean, coastal, and Great Lakes issues.

(2) Investment in a strong program of research, education, training, technology transfer, and public service is essential for this strategy.

(3) The expanding use and development of ocean, coastal, and Great Lakes resources resulting from growing coastal area populations and the increasing pressures on the coastal and Great Lakes environment challenge the ability of the United States to manage such resources wisely.

(4) The vitality of the Nation and the quality of life of its citizens depend increasingly on the understanding, assessment, development, utilization, and conservation of ocean, coastal, and Great Lakes resources. These resources supply food, energy, and minerals and contribute to human health, the quality of the environment, national security, and the enhancement of commerce.

(5) The understanding, assessment, development, utilization, and conservation of such resources require a broad commitment and an intense involvement on the part of the Federal Government in continuing partnership with State and local governments, private industry, universities, organizations, and individuals concerned with or affected by ocean, coastal, and Great Lakes resources.

(6) The National Oceanic and Atmospheric Administration, through the national sea grant college program, offers the most suitable locus and means for such commitment and involvement through the promotion of activities that will result in greater such understanding, assessment, development, utilization, and conservation. The most cost-effective way to promote such activities is through continued and increased Federal support of the establishment, development, and operation of programs and projects by sea grant colleges, sea grant institutes, and other institutions, including strong collaborations between Administration scientists and scientists at academic institutions.

(b) Objective

The objective of this subchapter is to increase the understanding, assessment, development, utilization, and conservation of the Nation's ocean, coastal, and Great Lakes resources by providing assistance to promote a strong educational base, responsive research and training activities, broad and prompt dissemination of knowledge and techniques, and multidisciplinary approaches to environmental problems.

(c) Purpose

It is the purpose of the Congress to achieve the objective of this subchapter by extending and strengthening the national sea grant program, initially established in 1966, to promote research, education, training, and advisory service activities in fields related to ocean, coastal, and Great Lakes resources.

Section 1122. Definitions

As used in this subchapter -

(1) The term ''Administration'' means the National Oceanic and Atmospheric Administration.

(2) The term ''Director'' means the Director of the national sea grant college program, appointed pursuant to section 1123(b) (FOOTNOTE 1) of this title.

(FOOTNOTE 1) See References in Text note below.

(3) the (FOOTNOTE 2) term ''director of a sea grant college'' means a person designated by his or her institution to direct a sea grant college or sea grant institute.

(FOOTNOTE 2) So in original. Probably should be capitalized.

(4) The term ''field related to ocean, coastal, and Great Lakes resources'' means any discipline or field, including marine affairs, resource management, technology, education, or science, which is concerned with or likely to improve the understanding, assessment, development, utilization, or conservation of ocean, coastal, or Great Lakes resources.

(5) The term ''institution'' means any public or private institution of higher education, institute, laboratory, or State or local agency.

(6) The term ''includes'' and variants thereof should be read as if the phrase ''but is not limited to'' were also set forth.

(7) The term ''ocean, coastal, and Great Lakes resources'' means the resources that are located in, derived from, or traceable to, the seabed, subsoil, and waters of -

(A) the coastal zone, as defined in section 1453(1) of title 16;

(B) the Great Lakes;

(C) Lake Champlain (to the extent that such resources have hydrological, biological, physical, or geological characteristics and problems similar or related to those of the Great Lakes);

(D) the territorial sea;

(E) the exclusive economic zone;

(F) the Outer Continental Shelf; and

(G) the high seas.

(8) The term ''resource'' means -

(A) living resources (including natural and cultured plant life, fish, shellfish, marine mammals, and wildlife);

(B) nonliving resources (including energy sources, minerals, and chemical substances);

(C) the habitat of a living resource, the coastal space, the ecosystems, the nutrient-rich areas, and the other components of the marine environment that contribute to or provide (or which are capable of contributing to or providing) recreational, scenic, esthetic, biological, habitational, commercial, economic, or conservation values; and

(D) man-made, tangible, intangible, actual, or potential resources.

(9) The term ''panel'' means the sea grant review panel established under section 1128 of this title.

(10) The term ''person'' means any individual; any public or private corporation, partnership, or other association or entity (including any sea grant college, sea grant institute or other institution); or any State, political subdivision of a State, or agency or officer thereof.

(11) The term ''project'' means any individually described activity in a field related to ocean, coastal, and Great Lakes resources involving research, education, training, or advisory services administered by a person with expertise in such a field.

(12) The term ''sea grant college'' means any institution, or any association or alliance of two or more such institutions, designated as such by the Secretary under section 1126 of this title.

(13) The term ''sea grant institute'' means any institution, or any association or alliance of two or more such institutions, designated as such by the Secretary under section 1126 of this title.

(14) The term ''sea grant program'' means a program of research and outreach which is administered by one or more sea grant colleges or sea grant institutes.

(15) The term ''Secretary'' means the Secretary of Commerce, acting through the Under Secretary of Commerce for Oceans and Atmosphere.

(16) The term ''State'' means any State of the United States, the District of Columbia, the Commonwealth of Puerto Rico, the Virgin Islands, Guam, American Samoa, the Commonwealth of the Mariana Islands, or any other territory or possession of the United States.

Section 1123. National sea grant college program

(a) Program maintenance

The Secretary shall maintain within the Administration a program to be known as the national sea grant college program. The national sea grant college program shall be administered by a national sea grant office within the Administration.

(b) Program elements

The national sea grant college program shall consist of the financial assistance and other activities authorized in this subchapter, and shall provide support for the following elements -

(1) sea grant programs which comprise a national sea grant college program network, including international projects conducted within such programs;

(2) administration of the national sea grant college program and this subchapter by the national sea grant office, the Administration, and the panel;

(3) the fellowship program under section 1127 of this title; and

(4) any national strategic investments in fields relating to ocean, coastal, and Great Lakes resources developed with the approval of the panel, the sea grant colleges, and the sea grant institutes.

(c) Responsibilities of Secretary

(1) The Secretary, in consultation with the panel, sea grant colleges, and sea grant institutes, shall develop at least every 4 years a strategic plan that establishes priorities for the national sea grant college program, provides an appropriately balanced response to local, regional, and national needs, and is reflective of integration with the relevant portions of the strategic plans of the Department of Commerce and of the Administration.

(2) Within 6 months of March 6, 1998, the Secretary, in consultation with the panel, sea grant colleges, and sea grant institutes, shall establish guidelines related to the activities and responsibilities of sea grant colleges and sea grant institutes. Such guidelines shall include requirements for the conduct of merit review by the sea grant colleges and sea grant institutes of proposals for grants and contracts to be awarded under section 1124 of this title, providing, at a minimum, for standardized documentation of such proposals and peer review of all research projects.

(3) The Secretary shall by regulation prescribe the qualifications required for designation of sea grant colleges and sea grant institutes under section 1126 of this title.

(4) To carry out the provisions of this subchapter, the Secretary may-

(A) appoint, assign the duties, transfer, and fix the compensation of such personnel as may be necessary, in accordance with civil service laws;

(B) make appointments with respect to temporary and intermittent services to the extent authorized by section 3109 of title 5;

(C) publish or arrange for the publication of, and otherwise disseminate, in cooperation with other offices and programs in the Administration and without regard to section 501 of title 44, any information of research, educational, training or other value in fields related to ocean, coastal, or Great Lakes resources;

(D) enter into contracts, cooperative agreements, and other transactions without regard to section 5 of title 41;

(E) notwithstanding section 1342 of title 31, accept donations and voluntary and uncompensated services;

(F) accept funds from other Federal departments and agencies, including agencies within the Administration, to pay for and add to grants made and contracts entered into by the Secretary; and

(G) promulgate such rules and regulations as may be necessary and appropriate.

(d) Director of National Sea Grant College Program

(1) The Secretary shall appoint, as the Director of the National Sea Grant College Program, a qualified individual who has appropriate administrative experience and knowledge or expertise in fields related to ocean, coastal, and Great Lakes resources. The Director shall be appointed and compensated, without regard to the provisions of title 5 governing appointments in the competitive service, at a rate payable under section 5376 of title 5.

(2) Subject to the supervision of the Secretary, the Director shall administer the national sea grant college program and oversee the operation of the national sea grant office. In addition to any other duty prescribed by law or assigned by the Secretary, the Director shall -

(A) facilitate and coordinate the development of a long-range strategic plan under subsection (c)(1) of this section;

(B) advise the Secretary with respect to the expertise and capabilities which are available within or through the national sea grant college program and encourage the use of such expertise and capabilities, on a cooperative or other basis, by other offices and activities within the Administration, and other Federal departments and agencies;

(C) advise the Secretary on the designation of sea grant colleges and sea grant institutes, and, if appropriate, on the termination or suspension of any such designation; and

(D) encourage the establishment and growth of sea grant programs, and cooperation and coordination with other Federal activities in fields related to ocean, coastal, and Great Lakes resources.

(3) With respect to sea grant colleges and sea grant institutes, the Director shall -

(A)(i) evaluate the performance of the programs of sea grant colleges and sea grant institutes, using the priorities, guidelines, and qualifications established by the Secretary under subsection (c), and determine which of the programs are the best managed and carry out the highest quality research, education, extension, and training activities; and

(ii) rate the programs according to their relative performance (as determined under clause (i) into no less than 5 categories, with each of the 2 best-performing categories containing no more than 25 percent of the programs;

(B) subject to the availability of appropriations, allocate funding among sea grant colleges and sea grant institutes so as to -

(i) promote healthy competition among sea grant colleges and institutes;

(ii) encourage successful implementation of sea grant programs;

(iii) to the maximum extent consistent with other provisions of this subchapter, provide a stable base of funding for sea grant colleges and institutes; and

(iv) encourage and promote coordination and cooperation between the research, education, and outreach programs of the Administration and those of academic institutions; and

(C) ensure compliance with the guidelines for merit review under subsection (c)(2) of this section.

Section 1124. Program or project grants and contracts

(a) Authorization; purposes; limitation on amount

The Secretary may make grants and enter into contracts under this subsection to assist any sea grant program or project if the Secretary finds that such program or project will -

(1) implement the objective set forth in section 1121(b) of this title; and

(2) be responsive to the needs or problems of individual States or regions.

The total amount paid pursuant to any such grant or contract may equal 66 2/3 percent, or any lesser percent, of the total cost of the sea grant program or project involved; except that this limitation shall not apply in the case of grants or contracts paid for with funds accepted by the Secretary under section 1123(d)(6) (FOOTNOTE 1) of this title.

(FOOTNOTE 1) See References in Text note below.

(b) Special grants; maximum amount; prerequisites

The Secretary may make special grants under this subsection to implement the objective set forth in section 1121(b) of this title. The amount of any such grant may equal 100 percent, or any lesser percent, of the total cost of the project involved. No grant may be made under this subsection unless the Secretary finds that -

(1) no reasonable means is available through which the applicant can meet the matching requirement for a grant under subsection (a) of this section;

(2) the probable benefit of such project outweighs the public interest in such matching requirement; and

(3) the same or equivalent benefit cannot be obtained through the award of a contract or grant under subsection (a) of this section.

The total amount which may be provided for grants under this subsection during any fiscal year shall not exceed an amount equal to 1 percent of the total funds appropriated for such year pursuant to section 1131 of this title.

(c) Eligibility and procedure

Any person may apply to the Secretary for a grant or contract under this section. Application shall be made in such form and manner, and with such content and other submissions, as the Secretary shall by regulation prescribe. The Secretary shall act upon each such application within 6 months after the date on which all required information is received.

(d) Terms and conditions

(1) Any grant made, or contract entered into, under this section shall be subject to the limitations and provisions set forth in paragraphs (2), (3), and (4) and to such other terms, conditions, and requirements as the Secretary deems necessary or appropriate. Terms, conditions, and requirements imposed by the Secretary under this paragraph shall minimize any requirement of prior Federal approval.

(2) No payment under any grant or contract under this section may be applied to -

(A) the purchase or rental of any land; or

(B) the purchase, rental, construction, preservation, or repair of any building, dock, or vessel;

except that payment under any such grant or contract may be applied to the short-term rental of buildings or facilities for meetings which are in direct support of any sea grant program or project and may, if approved by the Secretary, be applied to the purchase, rental, construction, preservation, or repair of non-self-propelled habitats, buoys, platforms, and other similar devices or structures, or to the rental of any research vessel which is used in direct support of activities under any sea grant program or project.

(3) The total amount which may be obligated for payment pursuant to grants made to, and contracts entered into with, persons under this section within any one State in any fiscal year shall not exceed an amount equal to 15 percent of the total funds appropriated for such year pursuant to section 1131 of this title.

(4) Any person who receives or utilizes any proceeds of any grant

or contract under this section shall keep such records as the Secretary shall by regulation prescribe as being necessary and appropriate to facilitate effective audit and evaluation, including records which fully disclose the amount and disposition by such recipient of such proceeds, the total cost of the program or project in connection with which such proceeds were used, and the amount, if any, of such cost which was provided through other sources. Such records shall be maintained for 3 years after the completion of such a program or project. The Secretary and the Comptroller General of the United States, or any of their duly authorized representatives, shall have access, for the purpose of audit and evaluation, to any books, documents, papers, and records of receipts which, in the opinion of the Secretary or of the Comptroller General, may be related or pertinent to such grants and contracts.

Section 1124a. Repealed. Pub. L. 105-160, Sec. 6, Mar. 6, 1998, 112 Stat. 24

Section 1125. Repealed. Pub. L. 102-186, Sec. 4(a), Dec. 4, 1991, 105 Stat. 1283

Section 1126. Sea grant colleges and sea grant institutes

(a) Designation

(1) A sea grant college or sea grant institute shall meet the following qualifications -

(A) have an existing broad base of competence in fields related to ocean, coastal, and Great Lakes resources;

(B) make a long-term commitment to the objective in section 1121(b) of this title, as determined by the Secretary;

(C) cooperate with other sea grant colleges and institutes and other persons to solve problems or meet needs relating to ocean, coastal, and Great Lakes resources;

(D) have received financial assistance under section 1124 of this title;

(E) be recognized for excellence in fields related to ocean, coastal, and Great Lakes resources (including marine resources management and science), as determined by the Secretary; and

(F) meet such other qualifications as the Secretary, in consultation with the panel, considers necessary or appropriate.

(2) The Secretary may designate an institution, or an association or alliance of two or more such institutions, as a sea grant college if the institution, association, or alliance -

(A) meets the qualifications in paragraph (1); and

(B) maintains a program of research, advisory services, training, and education in fields related to ocean, coastal, and Great Lakes resources.

(3) The Secretary may designate an institution, or an association or alliance of two or more such institutions, as a sea grant institute if the institution, association, or alliance -

(A) meets the qualifications in paragraph (1); and

(B) maintains a program which includes, at a minimum, research and advisory services.

(b) Existing designees

Any institution, or association or alliance of two or more such institutions, designated as a sea grant college or awarded institutional program status by the Director prior to March 6,

1998, shall not have to reapply for designation as a sea grant college or sea grant institute, respectively, after March 6, 1998, if the Director determines that the institution, or association or alliance of institutions, meets the qualifications in subsection (a) of this section.

(c) Suspension or termination of designation

The Secretary may, for cause and after an opportunity for hearing, suspend or terminate any designation under subsection (a) of this section.

(d) Duties

Subject to any regulations prescribed or guidelines established by the Secretary, it shall be the responsibility of each sea grant college and sea grant institute -

(1) to develop and implement, in consultation with the Secretary and the panel, a program that is consistent with the guidelines and priorities established under section 1123(c) of this title; and

(2) to conduct a merit review of all proposals for grants and contracts to be awarded under section 1124 of this title.

(e) Annual Report on Progress

(1) Report requirement.--The Secretary shall report annually to the Committee on Resources and the Committee on Science of the House of Representatives, and to the Committee on Commerce, Science, and Transportation of the Senate, on efforts and progress made by colleges, universities, institutions, associations, and alliances to become designated under this section as sea grant colleges or sea grant institutes, including efforts and progress made by sea grant institutes in being designated as sea grant colleges.

(2) Territories and freely associated states.--The report shall include descriptions of -

(A) efforts made by colleges, universities, associations, institutions, and alliances in United States territories and freely associated States to develop the expertise necessary to be designated as a sea grant college;

(B) the administrative, technical, and financial assistance provided by the Secretary to those entities seeking to be designated; and

(C) the additional actions or activities necessary for those entities to meet the qualifications for such designation under subsection (a)(1).

Section 1127. Fellowships

(a) In general

To carry out the educational and training objectives of this subchapter, the Secretary shall support a program of fellowships for qualified individuals at the graduate and post-graduate level. The fellowships shall be related to ocean, coastal, and Great Lakes resources and awarded pursuant to guidelines established by the Secretary. The Secretary shall strive to ensure equal access for minority and economically disadvantaged students to the program carried out under this subsection. Not later than 1 year after the date of the enactment of the National Sea Grant College Program Act Amendments of 2002, and every 2 years thereafter, the Secretary shall submit a report to Congress describing the efforts by the

Secretary to ensure equal access for minority and economically disadvantaged students to the program carried out under this subsection, and the results of such efforts.

(b) Dean John A. Knauss Marine Policy Fellowship

The Secretary may award marine policy fellowships to support the placement of individuals at the graduate level of education in fields related to ocean, coastal and Great Lakes resources in positions with the executive and legislative branches of the United States Government. A fellowship awarded under this subsection shall be for a period of not more than 1 year.

Section 1128. Sea grant review panel

(a) Establishment

There shall be established an independent committee to be known as the sea grant review panel.

(b) Duties

The panel shall advise the Secretary and the Director concerning -

(1) applications or proposals for, and performance under, grants and contracts awarded under section 1124 of this title;

(2) the sea grant fellowship program;

(3) the designation and operation of sea grant colleges and sea grant institutes, and the operation of sea grant programs;

(4) the formulation and application of the planning guidelines and priorities under section 1123(a) and (c)(1) of this title; and

(5) such other matters as the Secretary refers to the panel for review and advice.

The Secretary shall make available to the panel such information, personnel, and administrative services and assistance as it may reasonably require to carry out its duties.

(c) Membership, terms, and powers

(1) The panel shall consist of 15 voting members who shall be appointed by the Secretary. The Director and a director of a sea grant program who is elected by the various directors of sea grant programs shall serve as nonvoting members of the panel. Not less than 8 of the voting members of the panel shall be individuals who, by reason of knowledge, experience, or training, are especially qualified in one or more of the disciplines and fields included in marine science. The other voting members shall be individuals who, by reason of knowledge, experience, or training, are especially qualified in, or representative of, education, marine affairs and resource management, extension services, State government, industry, economics, planning, or any other activity which is appropriate to, and important for, any effort to enhance the understanding, assessment, development, utilization, or conservation of ocean, coastal, and Great Lakes resources. No individual is eligible to be a voting member of the panel if the individual is (A) the director of a sea grant college or sea grant institute; (B) an applicant for, or beneficiary (as determined by the Secretary) of, any grant or contract under section 1124 of this title; or (C) a full-time officer or employee of the United States.

(2) The term of office of a voting member of the panel shall be 3 years for a member appointed before the date of enactment of the National Sea Grant College Program Act Amendments of 2002, and 4

years for a member appointed or reappointed after the date of enactment of the National Sea Grant College Program Act Amendments of 2002. The Director may extend the term of office of a voting member of the panel appointed before the date of enactment of the National Sea Grant College Program Act Amendments of 2002 by up to 1 year. At least once each year, the Secretary shall publish a notice in the Federal Register soliciting nominations for membership on the panel.

(3) Any individual appointed to a partial or full term may be reappointed for one additional full term. A voting member may serve after the date of the expiration of the term of office for which appointed until his or her successor has taken office.

(4) The panel shall select one voting member to serve as the Chairman and another voting member to serve as the Vice Chairman. The Vice Chairman shall act as Chairman in the absence or incapacity of the Chairman.

(5) Voting members of the panel shall -

(A) receive compensation at a rate established by the Secretary, not to exceed the maximum daily rate payable under section 5376 of title 5, when actually engaged in the performance of duties for such panel; and

(B) be reimbursed for actual and reasonable expenses incurred in the performance of such duties.

(6) The panel shall meet on a biannual basis and, at any other time, at the call of the Chairman or upon the request of a majority of the voting members or of the Director.

(7) The panel may exercise such powers as are reasonably necessary in order to carry out its duties under subsection (b) of this section.

Section 1129. Interagency cooperation

Each department, agency, or other instrumentality of the Federal Government which is engaged in or concerned with, or which has authority over, matters relating to ocean, coastal, and Great Lakes resources -

(1) may, upon a written request from the Secretary, make available, on a reimbursable basis or otherwise any personnel (with their consent and without prejudice to their position and rating), service, or facility which the Secretary deems necessary to carry out any provision of this subchapter;

(2) shall, upon a written request from the Secretary, furnish any available data or other information which the Secretary deems necessary to carry out any provision of this subchapter; and

(3) shall cooperate with the Administration and duly authorized officials thereof.

Section 1130. Repealed. Pub. L. 102-186, Sec. 5(a), Dec. 4, 1991, 105 Stat. 1283

Section 1131. Authorization of appropriations

(a) Authorization--

(1) In general--There are authorized to be appropriated to the Secretary to carry out this title--
subchapter -

(A) $60,000,000 for fiscal year 2003;

(B) $75,000,000 for fiscal year 2004;
(C) $77,500,000 for fiscal year 2005;
(D) $80,000,000 for fiscal year 2006;
(E) $82,500,000 for fiscal year 2007; and
(F) $85,000,000 for fiscal year 2008.

(2) Priority activities.--in addition to the amounts authorized under paragraph (1), there are authorized to be appropriated for each of fiscal years 2003 through 2008--

(A) $5,000,000 for competitive grants for university research on the biology and control of zebra mussels and other important aquatic nonnative species;
(B) $5,000,000 for competitive grant for university research on oyster diseases, oyster restoration, and oyster-related human health risks;
(C) $5,000,000 for competitive grants for university research on the biology, prevention, and forecasting of harmful algal blooms, including Pfiesteria piscicida; and
(D) $3,000,000 for competitive grants for fishery extension activities conducted by sea grant colleges or sea grant institutes to enhance, and not supplant, existing core program funding.

(b) Limitations.--

(1)Administration.--There may not be used for administration of programs under this title in a fiscal year more than 5 percent of the lesser of--

(A) the amount authorized to be appropriated under this title for the fiscal year; or
(B) the amount appropriated under this title for the fiscal year.

(2) Use for other offices or programs.--Sums appropriated under the authority of subsection (a)(2) shall not be available for administration of this title by the National Sea Grant Office, for any other Administration or department program, or for any other administrative expenses.

(c) Distribution of Funds.--In any fiscal year in which the appropriations made under subsection (a)(1) exceed the amounts appropriated for fiscal year 2003 for the purposes described in such subsection, the Secretary shall distribute any excess amounts (except amounts used for the administration of the sea grant program) to any combination of the following:

(1) sea grant programs, according to their rating under section 204(d)(3)(A);
(2) national strategic investments authorized under section 204 (b)(4);
(3) a college, university, institution, association, or alliance for activities that are necessary for it to be designated as a sea grant college or sea grant institute
(4) a sea grant college or sea grant institute designated after the date of enactment of the National Sea Grant College Program Act Amendments of 2002 but not yet evaluated under section 204(d)(3)(A).

(d) Availability of sums

Sums appropriated pursuant to this section shall remain available until expended.

(e) Reversion of unobligated amounts

The amount of any grant, or portion of a grant, made to a person under any section of this subchapter that is not obligated by that person during the first fiscal year for which it was authorized to be obligated or during the next fiscal year thereafter shall revert to the Secretary. The Secretary shall add that reverted amount to the funds available for grants under the section for which the reverted amount was originally made available.

Section 9: COORDINATION

(This section has not yet been placed in the text sequence above.)
Not later than February 15 of each year, the Under Secretary of Commerce for Oceans and Atmosphere and the Director of the National Science Foundation shall jointly submit to the Committees on Resources and Science of the House of Representatives and the Committee on Commerce, Science, and Transportation of the Senate a report on how the oceans and coastal research activities of the National Oceanic and Atmospheric Administration, including the Coastal Ocean Program and the National Sea Grant College Program, and of the National Science Foundation will be coordinated during the fiscal year following the fiscal year in which the report is submitted. The report shall describe in detail any overlapping ocean and coastal research interests between the agencies and specify how such research interests will be pursued by the programs in a complementary manner.

Appendix I

Letter from James Coleman, Chair-NRC Committee Sent to All Individual Sea Grant Program Directors

THE NATIONAL ACADEMIES
Advisers to the Nation on Science, Engineering, and Medicine

Ocean Studies Board
500 Fifth Street, NW, Room 752
Washington, DC 20001
Phone: 202 334 2714
Fax: 202 334 2885
www.dels.nas.edu/osb

March 28, 2005

Dear *(director's name entered here):*

As Chair of the new National Academies committee to evaluate the Sea Grant review process I am writing to you to ask your assistance. In order to fully address our charge, we need information and materials to inform our process and give us input regarding the experiences of the state programs in the review process.

At the request of Congress, the Academies has named a committee to assess new procedures adopted since the release of the 1994 NRC report, and in particular examine such things as the effectiveness of those changes, the effectiveness of program review procedures, and an assessment of the usefulness and fairness of the metrics developed for program evaluation. The full task statement is available at: http://dels.nas.edu/osb/Seagrant.shtml.

I would like to request the following materials from your office:

- Any Final Evaluation Letters from NSGO
- PAT Reports from Cycle 1 and Cycle 2 (if available)
- Your response to the PAT reports
- Any Cycle 2 briefing materials you have available

In addition, I'd like you to consider the following questions, and send any thoughts or estimates you have to the Academies at the address listed above, to the attention of Ms. Nancy Caputo. Please understand that any materials that are shared with the committee will be included in a publicly accessible file. If you have concerns regarding material you deem sensitive please contact Jennifer Merrill, the study's director, at the phone number above to discuss possible alternatives.

What do you believe is the primary goal of the four-year review?

Do you feel your program has improved as a result of the PAT and evaluation process? If so, how?

Do you have any concerns about the current process you would care to share with the committee?

How would you change the current evaluation process, and why?

If you've completed Cycle 2 of the PAT process, do you have an estimate of the cost? Please try to give a breakdown of travel and meeting costs, publication, and staff time estimates. (Cycle 1 cost estimates were compiled by the SGA and the committee will be accessing those estimates.)

On behalf of the committee, I'd like to thank you for taking the opportunity to share your ideas and information with us. We are all looking forward to working with you and hearing from you in the coming year as we pursue our evaluation. If you should want additional information on the project or have questions regarding your submissions, please contact Jennifer Merrill at the address and phone given above, or at jmerrill@nas.edu.

Best regards,

James M. Coleman

Chair, NRC Committee to Evaluate the Sea Grant Program Review Process
Boyd Professor, Louisiana State University and Agricultural and Mechanical College

Appendix J

Executive Summary from "Review and Recommendations: Sea Grant Program Evaluation Process"

Draft Report
Dated November 18, 2005

Report of the Sea Grant Review Panel's
Program Evaluation Committee
Frank L. Kudrna, Chair
Peter Bell
Elbert (Joe) Friday
Manny Hernandez-Avila
Nathaniel E. Robinson
Jeffrey R. Stephan
Judith S. Weis

SUMMARY

Since the initiation in 1998 of the Program Assessment Team (PAT) concept, the National Sea Grant Review Panel (NSGRP) has been involved in the development, implementation and continuing evaluation of the PAT process. This began in 1997 when Carlos Fedderoff (representing the NSGRP), Bud Grisswald (formerly with the National Sea Grant Office (NSGO)), and B.J. Copeland (representing the Sea Grant Association [SGA]), developed the original recommendations concerning PATs into a report released July 30, 1997 entitled *Evaluation of Sea Grant College Programs: Recommendations for the Protocol, Criteria and Scheduling for Program Evaluation* (Disk Attachment).

An individual program undergoes a PAT review once during each 4-year PAT cycle. At the end of each year's PAT reviews, the NSGRP holds a "training session" in which the year's PATs are discussed and reviewed, and Panel members are trained to better evaluate programs. Topics such as strategic planning and metrics are examples of the material covered during this training. This document provides specific recommendations concerning the PAT Manual guidance and PAT training to improve future PATs.

After the completion of the first full cycle of PATs in 2001, a year-long review of the PAT process was conducted. This review process, chaired by Dr. John Toll of the Panel, produced the document *Review and Recommendations, Sea Grant Program Evaluation Process, Report of the Sea Grant Review's Panel Program Evaluation Committee, October 21, 2001,* otherwise referred to as the Toll Report (Disk Attachment). A significant number of these recommendations were accepted by the NSGO and incorporated in the second cycle of PATs that began in 2003.

At the midway point of the 2nd cycle of PATs, the National Sea Grant Review Panel charged its Program Evaluation Committee, at its November 2004 Sea Grant Review Panel meeting, making further recommendations concerning the PAT process. Frank Kudrna was appointed chairman of the Program Evaluation Committee, and the committee includes Peter Bell, Elbert (Joe) Friday, Manny Hernandez-Avila, Nat Robinson, Jeff Stephan, and Judy Weis.

Topics for consideration were requested from the Sea Grant Network. A formal response was sent by the SGA (Attachment 1). Support from a majority of the Committee was required for a topic to be developed into a white paper for consideration as a recommendation. The committee considered all of those topics recommended by members of the Sea Grant Network.

The committee established a format dividing its recommendations into three categories:

- Category 1—Recommendations concerning the current (second cycle) of PATs, to provide added guidance or clarification for the third year of the cycle.
  7 Recommendations.

- Category 2—Recommendations concerning the February NSGO Final Review.
  11 Recommendations.

- Category 3—Recommendations concerning the next (i.e., third) cycle of PATs.
  23 Recommendations.

All of the Category 1 recommendations (pages 1–2) were accepted and incorporated into the 2005 PAT Manual. Of the Category 2 recommendations (pages 2–4) concerning the February NSGO Final Review, the Director of the NSGO immediately implemented all but two of the recommendations. The two exceptions were held for later consideration (see Item 12, Category 3). The category 1 and 2 recommendations were presented to the Sea Grant Network, who were also asked for additional suggestions for Category 3 topics.

The Evaluation Committee held two conference calls to review all of the suggested Category 3 topics. Topics supported by a majority of the xommittee proceeded to a white paper, which included a description of the issue/problem, a discussion, and recommendations. These white papers were collected into a draft set of Category 3 eecommendations, which were distributed to the full NSRGP. Discussion of the draft recommendations occurred at the Wednesday Panel Training Session held during Sea Grant Week (Maine 2005). NSGRP members were given the opportunity to provide additional comments up to 6/15/2005. The Category 3 recommendations were then revised, and the Evaluation Committee held two conference calls to review drafts of the recommendations.

We are extremely pleased that this document is the consensus document developed by the xommittee with no minority views, by the Executive Committee, approved by the full Sea Grant Panel.

Lastly, although our xommittee has made a series of recommendations to improve the National Sea Grant College PAT process, we must comment on the high quality of the existing PAT process. The Sea Grant College Programs have been given an enormous amount of latitude to run, direct, and provide matching funding for their programs, and are retrospectively reviewed by the PAT process. We believe the Sea Grant PAT model is the most rigorous and comprehensive evaluation process to be found anywhere. Ron Baird, the Director of the National Sea Grant Office, has been recognized by NOAA for his leadership in developing this program through the Presidential Rank Award for Meritorious Service, and an adaptation of the PAT review process is now utilized by the National Institute of Health in their reviews. We believe the recommendations contained in this report will make an excellent review process even better.